Bellamy on Botany

Bellamy on Botany

by

David Bellamy

BRITISH BROADCASTING CORPORATION

The programmes were produced by
David Cordingley and Michal Weatherley

The television programmes
First broadcast on BBC-1,
April to June 1972

Acknowledgement is due to the following for permission to reproduce photographs:

AEROFILMS LTD plate 26; HEATHER ANGEL plates 14 and 27;
DR DAVID BELLAMY plates 22 and 23; G. B. BURGESS plates 8 and 9;
JOHN MARKHAM plate 20; NATURAL HISTORY PHOTOGRAPHIC AGENCY (photo
Stephen Dalton) plate 21; T. PETERS plates 4, 5, 6 and 7; J. REDHEAD front
cover and plate 10; RENTOKIL LABORATORIES LTD plate 19; ROYAL BOTANIC
GARDEN, EDINBURGH plates 24 and 25; STATE INSTITUTE OF AGRICULTURE OF
THE NETHERLANDS plates 11, 12 and 13; WELSH PLANT BREEDING STATION
plates 1, 2 and 3.

The maps on pages 68 and 69 were reproduced by permission of F. H. PERRING
and S. M. WALTERS in *Atlas of British Flora* published by Thomas Nelson &
Sons, Ltd.

The drawings are by DAVID COOK and the diagrams are by CHARLES
MATHESON. Diagrams in the Do it yourself Botany Kit are by
PETER TAYLOR.

Published by the British Broadcasting Corporation
35 Marylebone High Street, London W1M 4AA
This book is set in Monotype 11/12pt Garamond
Printed in Great Britain by Lowe & Brydone (Printers) Ltd, Thetford,
Norfolk

ISBN 0 563 10666 2

Contents

	Preface	vii
	Introduction	I
1	Carry on Cutting	2
2	In the Mire	8
3	Deeper in the Mire	18
4	The Kingdom of Canute	24
5	You Can't See the Wood	32
6	Decay and Delicacy	39
7	Where have all the trees gone?	48
8	In the Swim	58
9	The Wars of the Primroses	67
10	What is Vegetation?	74
	Do it yourself Botany Kit	81

TO 'HUTCH'

Preface

This revised edition of 'Bellamy on Botany' has resulted from the popularity of the original publication and the television programmes which it supplemented. Many viewers were prompted to enquire how they might actively pursue their interest in natural history. David Bellamy has now added a set of supplementary notes and suggestions for activity in the hope that this demand will be satisfied. With infectious enthusiasm and humour he gives a fascinating insight into botany. He does not just see plants as interesting curiosities, but as the end products of millions of years of evolution on whose existence we depend and which deserve our respect.

By understanding how the plant communities have developed, it is possible for us to help to control and preserve our environment in the future.

Introduction

Bellamy on Botany was not my idea, I wanted to call it, 'Why do we take plants for granted?' 'No', groaned the producer, and as he had said those nice things about me in the preface, I let him have his own way.

It certainly is true that we take plants for granted, and it is also true that the usual idea of a botanist is of someone grubbing about in the grass with a magnifying glass, and pressing flowers. I am not denying that botanists do such things, but there is much more to botany than that. Today botany uses the whole range of sophisticated scientific techniques, computers, mass spectographs, electron microscopes, etc. – you name it, we have it! This subject is in the forefront of modern scientific exploration and is seeking answers to some of the key problems facing the future of mankind.

To make a selection from such a vast subject is a problem. In the event I decided to give you a very personal selection of some of the things which I find particularly fascinating in the world of plants. What I hope I have done is to show why I enjoy being a botanist and that botany is not just a subject for the expert. There is a lot to see and a great many conclusions to be drawn by anyone prepared to keep his eyes open as he goes around the countryside.

The fact that two of the ten chapters (and television programmes) are devoted to peat bogs will probably tell you, correctly, that this is my own particular interest. I am certain you are in for some surprises in the squelching world of the peat ecologist, and I hope you will look with a new interest at sand dunes, forests, primroses, even such a common thing as grass, and the other subjects of this book.

DAVID BELLAMY

1 *Carry on Cutting*

Mowing the lawn is almost an obsession with the English, and very few of us give much thought to exactly what we are mowing.

It's grass down there, and grass is pretty important stuff, without it the world would be a very different place. From the tundra to the tropics, from the edge of the sea to the tops of our mountains, on the open plains, in the depths of the forest, just about wherever there is life there is grass. The open sea is the only habitat on earth which the highly successful grass family, the GRAMINEAE, has not evolved to exploit. What is it then that makes grass such a success?

Part of the answer is the reason we mow the lawn. If you chop the top off most plants they either die, or take a long, long time to recover. Not so grass – in fact, if it's the stuff that's on my lawn, it seems to thrive on it.

The real difference is that the main part of a grass plant consists of leaves. Go and get hold of one and strip the leaves off (see fig. 1) and have a real close look. Each leaf arises from a node on the stem (or culm) and then tightly encircles the stem. This is the leaf sheath from which the lamina arises. The leaf has an intercalary meristem (new tissue is laid down from near the base), and if you chop the top off the leaf, it still goes on growing.

The most important part of an ordinary plant is the stem which has an apical meristem (it grows from the top), producing new leaves as it goes. If you chop the top off, it will take a long time for growth to get going again.

Now you know why you have to mow the lawn, but the grasses did not wait for the invention of the lawnmower; they must have evolved in order to survive alongside the grazing animals. This demonstrates that the whole of evolution is an integrated process.

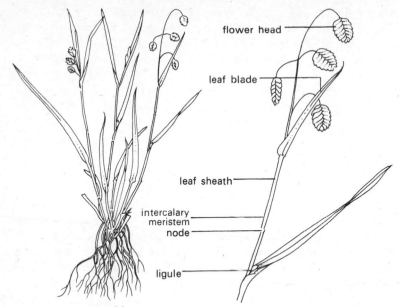

Fig. 1. The great quaking grass.

Grazing animals need a reliable source of herbage and the grasses evolved under the pressure of grazing, seizing the opportunity of the range.

The other reason for the success of the grasses is their flower. Yes, grasses do have flowers and some of them, like the Great Quaking grass, are very beautiful. Fig. 2 shows a grass flower and an ordinary flower partly dissected to show the parts. Flowers are

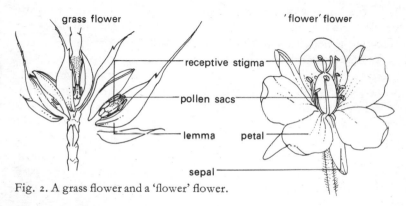

Fig. 2. A grass flower and a 'flower' flower.

simply shoots which have been modified for reproduction, the key task of the flower being to disseminate its pollen and receive pollen from another flower. The very showy flowers with gorgeous scents rely on insects to carry the pollen, and this of course can be a chancy process, especially in these days of D.D.T. The grass with its scentless, rather dull flower relies on the wind to carry the pollen, and of course there's a good supply of wind out in the open grasslands, and pollination is very efficient. That is one reason for the dreaded hay fever. (It must be terrible to be a cow, or for that matter a botanist, and suffer from hay fever.)

Take a long cool think. What would the world be like if the GRAMINEAE had never evolved, what would life be like without grass? There wouldn't be much life; take ourselves as an example, we rely absolutely on this one family of flowering plants. Without rice the staple diet of two-thirds of the world's population would be gone, no bread or macaroni – this takes care of the other third, and even Marie Antoinette would not have been able to enjoy her cake! Chips with everything would be the order of the day, potatoes the staple diet. There would be no corn oil, so it would be back to animal fats for frying, but what would the animals eat? Without the grass fields there would be no permanent cover to the soil, so once the forests had gone, soil erosion would be widespread and disastrous. If Marram grass had not built the sea defences the whole shape of our coastline would be changed, much of Holland a mere memory, and great tracts of now fertile lowlands covered with drifting sand. The only good thing about life without grass would be no mowing the lawn, but also no beer or whisky with which to while away the leisure hours.

In the same way that the grasses evolved alongside the great grazing animals, man's agrarian society has evolved around grass, reaping the benefit of this great family of flowering plants.

At first nomadic man in his wanderings began to learn about the GRAMINEAE, which species were useful, which were not; some could grow in hot dry places, others in cool wet places. Selecting from this knowledge, the early agriculturalists tilled the ground and sowed their new-found crops. Thus, wandering hunters became fixed agriculturalists. The first successful experiments in food production were probably in the fertile crescent of south-western Asia, and only about 9,000 years ago. The two grasses, or rather cereals, domesticated were wild wheat, and wild barley. It is rather interesting to speculate on what exactly happened. Gathering, say wild barley, would have presented an immediate problem because the ripe seed head is very easily broken up, releasing grain, so the obvious thing would be to collect the ones that were less brittle. These would be carried home to the cooking pot, or stored to be used as seeds for a new crop. So even then selection of a more useful type of barley would be taking place. There is good evidence of these early crop plants because not only have preserved food stores been found but so has pottery, which can be accurately dated, decorated with impressions of the cereal heads. It is in this way we know that although the barley used by early agriculturalists had six rows of flowers and hence grains on each head, the original domesticated stuff had only two rows on the head, a much less productive plant.

So it must have gone on, more and more knowledge allowing better selection of better plants for the next year's crop. Then very recently, with a knowledge of the pollination mechanism of the cereals, and of the laws of heredity, the possibility of successfully breeding the right crop plant for the right job became reality.

One very important feature of the cereals is that they are self-pollinated. This simply means that the pollen from the anthers of one flower pollinates the stigma of that flower. Now hereby hangs a tale because basically the same genetic characters will be mixed each time and a cereal left on its own will carry on breeding more or less true to form every time. So if you want to bring about cross pollination, that is to transfer the pollen, and the characters from the anthers of one flower to the stigma of another, you must go through the whole palaver of emasculating the recipient by removing the developing pollen sacks and then bringing the pollen from another plant to the stigma. It is a somewhat tedious process, but the results can be more than worthwhile.

Take for instance the cross between wheat (*Triticum*), as the father, and rye (*Secale*), as the mother, the new offspring is *Triticale* (plates 1, 2, 3). But look at it – a super cereal, large flower head full of grain and full of promise.

This simple pattern of selecting plants with the required characteristics (high yield, tolerance of prevailing climatic conditions, low susceptibility to fungal diseases, etc.) and then interbreeding them by the method of cross-pollination sounds easy, but the permutations and combinations are enormous and form the meat of

Plates 1, 2, 3. Left, Rye *Secale* (dad), right, wheat *Triticum* (mum), with their offspring *Triticale* – a potential super cereal.

the exacting science of plant breeding. World research centres, like the International Maize and Wheat Improvement Centre, the International Rice Research Institute and the Welsh Plant Breeding Station, study breeding within the royal family of the grasses and cereals, and the geneology of each ruling monarch is known in great detail. Plants with pedigrees may sound a completely closed shop situation. However, it is fascinating to think that next time you go on holiday, you could come across a wild grass, or cereal, with just the right characteristics needed in some high society set-up – the commoner who could marry into royal circles. The only problem is that you must know what you are looking for.

Put yourself in the place of a grass on a modern farm. High stock density, that is lots of cows in the field, means that you must be able to tolerate not only excessive eating, but also excessive trampling and manuring. What a life! But the grass has to put up with it, and one of the programmes of research going on at the Welsh Plant Breeding Station is to find the perfect pasture plant, productive and always cheerful under the most appalling conditions. As a botanist I must protest, why not breed cows with better manners? 'Grasses of the world unite' could be the battle cry of the Green Revolution. However, the real Green Revolution is being masterminded by the plant breeders working to produce the grasses and cereals vital to this starving world.

2 *In the Mire*

You can burn it, make paper out of it, put it in your garden, or, if you are sufficiently rheumatic, take a bath in it. Whichever way you take it, peat is pretty useful stuff. But what is peat and how is it formed? The first prerequisite is stagnant or slow-moving water. Now if leaves or other vegetation fall into this water, they will not rot, or if they do, it will only happen very slowly, because decomposition depends on oxygen and the gas can only dissolve and diffuse through the water very slowly.

Take as an example a shallow lake, fed by a small stream which drains a hilly landscape. The stream not only forms the water supply to the lake, but also brings with it an abundance of the minerals which are necessary for the healthy growth of plants.

A profusion of water plants begins to grow in the lake. First, only free-floating plants can get a foothold, plants such as the microscopic planktonic algae and the duckweed (see chapter 8). However, quite soon, and for obvious reasons, a distinct zonation of types of plants comes into evidence around the lake. The members of each zone are adapted to the depth of water in which they are growing (fig. 3).

However, the zonation is not a static thing, each of the zones is on the march because the annual fall of plant litter, leaves, stems, twigs, which accumulate in the form of sapropel and peat, slowly fills the lake starting at the shallowest part gradually working forward.

So a body of water can in time become no more than a beautiful memory replaced by an equally beautiful natural woodland.

When we were filming these programmes, we visited one of these lakes in Shropshire. I hadn't been there for some years and

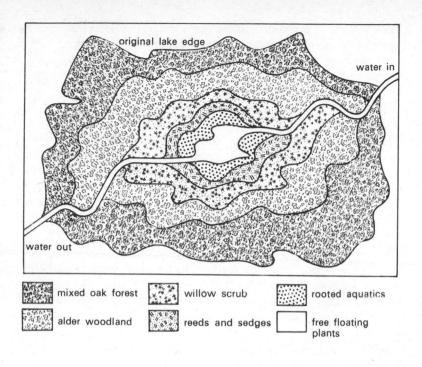

mixed oak forest   willow scrub   rooted aquatics

alder woodland   reeds and sedges   free floating plants

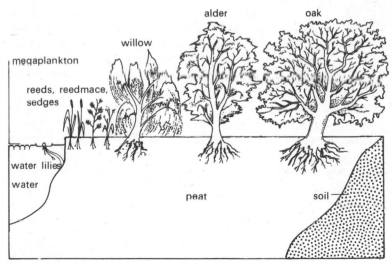

Fig. 3. Zonation around a vanishing lake, in plan above, and profile.

on the last visit I had swum in the open water. It's quite staggering to see that in only this short period, the same spot had been swallowed up by the encroaching vegetation. Apart from a few patches of dark peaty water surrounded by thick black, peaty mud, the lake is just a memory (plate 4).

If you had the patience (and the longevity) to sit in the middle of one of these lakes for a very long period and record what happens in a notebook, you could trace the development for yourself. Alternatively, you could invest in a peat borer, a simple machine (fig. 4), which can be pushed down into the peat to collect samples at regular intervals. The whole process of colonisation can then be laid out before your very eyes, and using either the technique of pollen analysis, or radiocarbon dating (see chapter 3), an accurate time-scale may be determined.

Yes, every peat deposit is a history book, and anyone willing to use their eyes can begin to read it. Remember, peat is partially decayed organic matter and some parts of plants and animals decay very slowly. It may take an expert to identify the tiny pollen grains which are preserved in the peat record but most people with only a little knowledge of natural history are capable of recognizing things like alder cones, sedge fruits, the rhizomes of the common reed, etc. (fig. 5).

Plants grow producing stems, leaves and roots, and the process of growth requires energy – energy from the metabolism of chemical substances. All plants metabolize and all plants produce acidic waste as a by-product of these metabolic processes. In an open lake, like the one described, the acidic substances will be neutralised by bases carried by the stream water, and will be carried away out of the lake by the flow of water.

However, in certain situations this is not the case and the acid substances begin to accumulate. If this happens the type of peat and the plants growing on it, and subsequently forming it, begin to change, and this is exactly what happens in a lake basin that has neither inflow nor outflow streams. Basins of this type are common in landscapes which during the last ice age were situated at the meeting point of two glaciers. It is not difficult to imagine how the enormous pressures set up in such a situation could push great blocks of ice deep into the underlying drift material of sand and silt. As the glacier moved on or melted, these pockets of dead ice were left behind and eventually even the dead ice would melt,

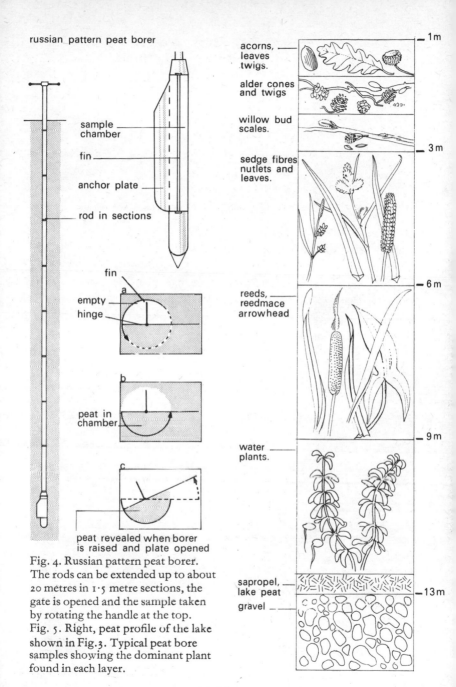

russian pattern peat borer

sample chamber

fin

anchor plate

rod in sections

fin

a

empty

hinge

b

peat in chamber

c

peat revealed when borer is raised and plate opened

acorns, leaves twigs.

alder cones and twigs

willow bud scales.

sedge fibres nutlets and leaves.

reeds, reedmace arrowhead

water plants.

sapropel, lake peat

gravel

— 1 m

— 3 m

— 6 m

— 9 m

— 13 m

Fig. 4. Russian pattern peat borer. The rods can be extended up to about 20 metres in 1·5 metre sections, the gate is opened and the sample taken by rotating the handle at the top.
Fig. 5. Right, peat profile of the lake shown in Fig.3. Typical peat bore samples showing the dominant plant found in each layer.

Plate 4. General view of an open lake in the final stage of the disappearing act. An autumn view of the lake.

Plate 5. The original gonk, tussocks of the Great Panicled Sedge protruding from the black, peaty mud.

Plate 6. Reeds and other water plants whose dead remains will continue the vanishing act.

Plate 7. The edge of the advancing wood.

producing lakes with no inflow or outflow stream. Basins formed in this way are called kettle holes and deep, steep-sided lakes of this type are common in some areas. In such lakes a different and surprising kind of peat formation can occur.

There are no shallow margins in such a lake (fig. 6) and the only type of plants which can grow are either the free-floating ones, or those which anchor at the edge and quickly grow out across the surface of the water. Furthermore, the water is stagnant and so all the acid by-products of the metabolism of the plants will accumulate. These acid stagnant conditions are ideal for the growth of the bog moss SPHAGNUM.

Sphagnum has two very important features. First and foremost it has the property of being able to grab hold of just about any nutrient which comes its way; in fact, too many nutrients such as potassium, calcium and even nitrate in the water will kill it off. Secondly, the plant consists of great masses of leaves, each of which is made up of a honeycomb of living and dead cells (see chapter 3, plate 10). The dead ones have pores and can fill with water, the whole thing acting like a sponge.

Sphagnum soon gets a foothold and, together with a highly adapted group of plants, begins to form a skin of living peat out across the surface of the water, gradually covering the lake. The remnant of open water, the only visible sign of the dying lake becomes smaller and smaller until it disappears. These black mysterious pools in the often vividly coloured skin of living peat look very much like great sightless eyes and in the German language are called MOOR EYES.

Long before the moor eyes are finally closed, a curious freak of nature occurs (plates 8 and 9). Pine trees begin to colonize the living skin of peat. The youngest and smallest trees are always found nearest the edge of the pool, the largest and oldest on the thicker more consolidated peat near the edge of the basin. The floating forest has a concave skyline dipping down to the dying lake in the middle.

So it is possible to have the uncanny experience of standing in a mature pine forest suspended by just a few feet of wet peat and roots over very deep water. What is more, you can be more than a kilometre from terra firma. The German language also contains a word for this terra infirma and it is very descriptive. The word is SCHWINGMOORE because, one presumes, it SCHWINGS about when you walk on it.

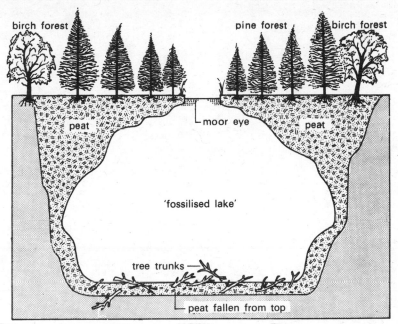

Fig. 6. A section of a kettle-hole lake, almost covered by a floating forest of Pine. Note the Birch trees at the edge where there is some influence of nutrients draining in from the catchment.

Again there are two ways of proving that you are not standing in an ordinary forest. First, use the peat borer and push it into the surface; it is tough going down through the tangle of roots, through the living skin, but then SLLURPP it drops like a stone into the hidden lake. 'More good peat borers have been lost that way!'

The second way is simple – just jump up and down. Very soon the trees are swaying wildly far above your head, riding high on the great waves which pass across the schwingmoore.

It is possible, with the right gear, and adequate training and precautions, to scuba dive underneath the skin. Of course, it is pitch dark down there and even a light does not help much because you are swimming in a thick soup of suspended peat and sphagnum leaves. Deeper, however, the water is clear but, of course, there is nothing to see although occasionally one may make sudden and dramatic contact with a water-logged tree trunk that has fallen

through from above. So, as long as you can put up with very cold sphagnum soup, garnished with an unmistakable taste of hydrogen sulphide, you too can have the experience of swimming in a fossilised lake. Yes, believe it or not, there are still places even in overcrowded England, as yet unexplored by man.

Once submerged, if you are lucky enough to see anything, you will find that the skin is, in part, supported by a great complex mass of the roots and stems of marsh plants. Two of the most abundant plants which play this supporting role are the Marsh Cinquefoil (*Potentilla palustris*), and the Bog Bean (*Menyanthes trifoliata*).

In a landscape the vanishing lakes are living history books and very little can happen in the vicinity of a growing peat deposit that will not be recorded in detail for posterity. All that posterity needs is a peat borer and a tame palynologist – someone who can recognise pollen (see chapter 3).

If you do decide to invest in a peat borer, don't be surprised by what you find in a peat bog. Marsh gas is very commonly encountered bubbling up around the borer. This can be lit to produce a sizeable flame leaping from the ground beneath your feet. This, together with phosphene, is a by-product of the processes of slow decay. Phosphene is self-igniting and is, of course, the Will o' the Wisp light which is said to have led many people to their deaths, lost in the schwingmoore. One peat bog, and I am not going to let on where it is, sits in a funnel of blue volcanic clay and, with a bit of luck, the final peat core will contain diamonds – and that is no fairy tale; so it's well worth getting into the mire.

Plates 8 and 9. General view of a kettle-hole lake in the final stage of the disappearing act. That forest is floating over 50 feet of water. In the foreground a lawn of Sphagnum moss with bog cotton at the edge of the lake.

3 *Deeper in the Mire*

If you are among the select few whose brief existence on this earth will be recorded for posterity in granite, or some other substance not easily decomposed, don't read any further. However, for all those in a less exalted position and who crave to leave behind some permanent record, then follow these instructions. Place a clause in your will to the effect that your body must be interred in a shallow grave in an acid peat bog. This will guarantee your own very personal record, perfectly preserved even down to your finger prints, for at least 2,000 years. Certainly this is the length of time for which Tollund man lay perfectly preserved in his peaty retreat, and the only reason that he isn't still there is that someone dug him up. So choose your peat bog, and choose it well. The fact is that nearly everything that falls onto a growing peat bog will be preserved and kept safe in the sub-fossil record of the developing peat deposit.

Every year all plants produce either spores or pollen grains of some sort and many of these have, for the expert, characteristics which are identifiable under the microscope (fig. 7).

Anyone who suffers from hay fever will understand that pollen grains do have specific characteristics. Without any expert training our nasal membranes can make very accurate determinations of the type of pollen grain floating about in the air. These spores and pollen grains fall onto the surface of a peat bog and become preserved to form an historical record of what has happened to the landscape surrounding the peat bog. Peat bogs are therefore active and accurate chronicles of the past, real living history books, because as the peat bog grows upwards the record each year is

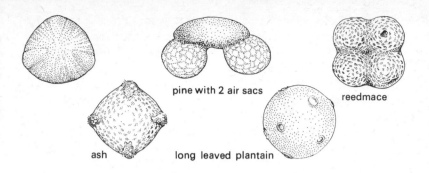

pine with 2 air sacs

reedmace

ash

long leaved plantain

Fig. 7. Pollen grains – a palynologist's view down the microscope.

preserved in the exact sequence of time. So get out the peat borer, find the right peat deposit and with luck all is revealed.

The first part of the process is very mucky and the last part exacting. Down goes the borer and up comes the column of peat, dark brown or black and often very sticky. Once exposed, sub-samples are taken and stored in closed containers, taking care to avoid contamination with the contemporary spores in the air. Each sample must be digested with strong acids to remove all superfluous inorganic and organic material. The resultant, after concentration in a centrifuge (a very fast, highly sophisticated spin dryer), is stained to help bring out the characteristics of the spores and grains and can then be studied under the microscope. Even greater care must be taken if the samples are to be used for radio-carbon dating. This is a technique which allows some organic matter to be dated with an accuracy of give or take 100 years. It depends on the fact that due to cosmic radiation a certain amount of the carbon dioxide in the atmosphere contains the radioactive isotope C^{14} which, like all radio-isotopes, gradually decays. Knowing the rate of decay of the isotope and the amount of radioactivity still left in the sample (this is the difficult bit that can easily get fouled up by contamination), it is possible to estimate when the organic matter was formed.

Identifying and counting all the different sorts of spores and pollen grains in a sample requires much expertise and even more patience, but this does tell what plants were present and very roughly how important each one was in the landscape. Immediately you can see that there will be a lot of problems. Do all types preserve equally well? Are all pollen and spore types transported equally effectively? For example, the pollen of the pine tree (fig. 7) which has great expanded air sacs, or of a tree like the hazel, with its super pollen factories, the catkins, are ideal for wind transport and are almost certain to be detected in a peat sample; whereas the orchid, which relies on insects for transport, has a much lower chance of being found.

However, the pollen people (Palynologists) have been doing their homework and have schemes which account for just about everything, and so the painstaking counts can be translated into the wonderful story of how we got, not only our flora, but also our vegetation.

The diagram (fig. 8) shows a simplified picture of what happened over the last 10,000 years in a typical part of lowland England. We can immediately infer the major climatic changes from the vegetation changes; there has been an overall warming up since the last ice age. The open tundra gave way gradually to park tundra with birches, and then, as the main pulse of migrations got underway, a landscape swathed in forest developed. Imagine the problems of travel through a wholly wooded countryside, with no tracks, no bridges and the natural drainage producing vast tracts of lowland swamp. Yet it was into this rich productive landscape that Neolithic man first came. Man, the hunter, reaped the benefits of the forest and soon left his own personal mark. The effect was dramatic, a drop in the number of pollen grains derived from trees and an increase in the pollen of herbs, grasses and especially of the things that we label today as weeds. Whether man was entirely responsible for the destruction of the forest is still in doubt because about the time that man arrived on the scene in force, the weather was beginning to deteriorate. Man, first with stone axes and later with the aid of iron, slowly destroyed the lowland forests producing arable land and pasture. In the uplands, man and the effect of the much wetter climate replaced the forest by the great blanket of peat which even today gives our mountain landscape its own special appearance.

Blanket Bog is the most extensive of the living history books, covering the whole landscape and noting down everything man and nature does within that landscape. One plant which plays a vital role in the formation of the bogs, and hence becomes their chronicler, is the bog moss, *Sphagnum imbricatum*, a very distinctive

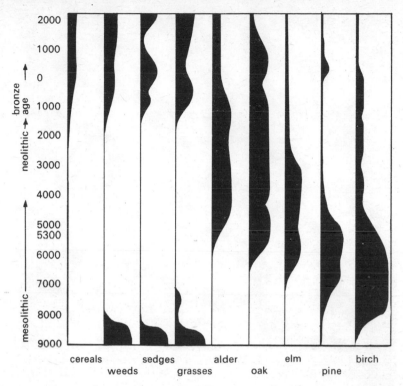

Fig. 8. A simple pollen diagram from a peat deposit in the lowlands of England. The diagram is an historical account of the vegetation of the landscape since the glaciers of the last glaciation melted, an historical record of more than 10,000 years. The record was dated by pollen and radiocarbon content.

plant with its conspicuous comb fibrils (plate 10). We know that at one time this particular bog moss was both common and abundant over large tracts of our uplands. Today it is a rare plant. The reason, in all probability, is due to the burning cycle which is the basic form of upland management. Burning old heather, which grows with Sphagnum, provides young shoots; food for grouse and sheep alike. This practice, especially where coupled with the construction of artificial drains has killed the living peat blanket and has initiated massive erosion.

The statistics of the world peat resource are astounding, there are 250 million acres of land covered with peat which weighs 223

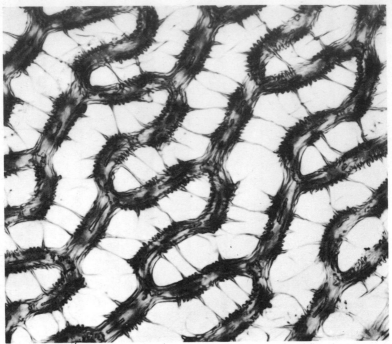

Plate 10. Close-up of the leaf of the bog moss *Sphagnum imbricatum*. The leaf consists of a sponge-like mass of dead cells, each of which has large open pores, and narrow-living cells containing chlorophyll.

billion tons, and holds 650,000 billion gallons of water. It has many uses, from the making of whisky to the great chemical industries which are based on peat wax. These industries can also get their power from peat-fired power stations. Russia alone has seventy-seven such power stations, many having capacities of over 300 Megawatts.

The world's resource is being used at an ever-increasing rate and the virgin mires are being drained and reclaimed for agriculture and forestry. The end result of both processes is the same, the organic matter is oxidised away producing vast amounts of carbon dioxide. If all the peat in the world were oxidised away, it would produce 3,000 billion tons of CO_2 which would pass into the atmosphere. Now CO_2 absorbs the heat rays from the sun (the infra red part of the sun's radiant energy) and if all the CO_2 did get into the atmosphere and stayed there, the temperature of the earth would increase, so much so that all the glaciers and ice caps would melt and the sea level would rise by something over 100 feet. All of man's main centres of industry, commerce and agriculture would be under water, and man would have to move onto the uplands. However, he could do it in the sure knowledge that there was no peat bog taking down the evidence.

4 *The Kingdom of Canute*

If you 'do like to be by the seaside' it is probably for one of two reasons, sea or sand. A sandy beach lapped by a warm sea is a pretty idyllic place to idle away a few hours, but think how nasty it is when the wind gets up. Try eating your picnic with a crisp north-easter blowing; a dream becomes a nightmare and sandwiches become sandwiches with a vengeance. Now think of all the problems there would be if you had to live there all the time. Think about your favourite sandy beach, it is very clear that apart from the summer visitors, precious little does live there. But one plant has become adapted to live in this hostile environment, *Ammophila arenaria*, the Marram grass.

Sand is finely ground rock. Anywhere there is a supply of sand, or readily erodable sandstone in the sea, and prevailing onshore winds, there is going to be a problem with sand moving inshore. If evolution had not produced the Marram grass, or something like it, then very large tracts of our lowlands would be swamped by a sea of shifting sand, and other areas would be swamped by the sea itself.

Sand dunes are nature's own sea defences and in certain parts of the world there are dune systems which are over 1,000 feet high. They don't get that big in Europe, but all are of great importance, such as the ones along the coast of Holland.

Anything which either grows or becomes deposited on the beach can form a nucleus for dune formation. Those ghastly annoying flies that flit about the beach and spoil your sunbathing give the whole game away. Next time you are on a beach and the wind is blowing, stand a matchstick upright in the sand. Immediately any flies in the vicinity will come and perch on the lee side of the matchstick and a miniature dune will begin to form on the same side.

When the embryo dune has become large enough, the first plant to become established is Sand Couch grass. This is very closely related to the common Couch grass, which often is the bane of the life of gardeners and farmers alike. The Couch grasses have extensive creeping underground systems of stems and roots. (It's easy to identify a stem, it has nodes, internodes and leaves. If it occurs underground, the leaves are usually reduced to mere scales and then the stem is called a rhizome.) The rhizomes and roots bind the sand together, bringing stability to the process of dune formation. Soon however the Sea Couch grass is replaced by the Marram, the root and rhizome system of which is even more branched and penetrates the sand very deeply. Now neither of these grasses produce extensive root systems just to hold the developing dune together, the roots grow into the sand in search of water. Sand is very porous stuff and any rain drains straight through the loosely packed surface layers. Down go the roots, in search of water, solving one problem but creating another. The sand builds up so rapidly that the Marram grass would be buried but for the fact that new shoots are being continually produced at the top, thus keeping its head above the sands of time.

The real success story of the Marram lies in its leaf (fig. 9 on the following page). The leaves have evolved to carry out the process of photosynthesis and there are three raw materials basic to the process, light from up above, water up from the roots, and carbon dioxide which must diffuse into the leaves from the air. So all leaves, except those which grow under water have tiny pores, the stomata (see chapter 8), which let the carbon dioxide in; the only trouble is that they also let the precious water out as water vapour. All land plants therefore face a dilemma; enough carbon dioxide must get in without letting too much water get out. To a plant like the Marram, growing in that sandy desert, it really is a big problem. Evolution has given this grass leaves which can roll up, rather as some of us can roll our tongues. When it rolls up, all the stomatal pores, through which the main loss of water takes place, are rolled up inside. As a further defence they are covered with a dense weft of hairs and so very little water can escape. The Marram grass can therefore grow on the inhospitable sand,

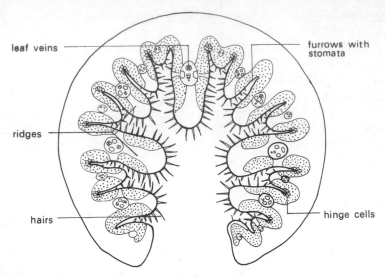

leaf veins

furrows with stomata

ridges

hairs

hinge cells

Fig. 9. Section across the rolling leaf of Marram.

causing the build-up of the dunes; a real case of 'rolling leaves will gather lots of sand', in fact 'will gather lots of moss' as you will see later.

The young sand dune with its Marram grass is one of the few natural monocultures in the world. The grass not only extracts water from the dune but it also adds humus and the surface of the sand begins to change. This is very obvious because the younger dunes near the sea look white; the pure white of the sand. As they age, the surface sand slowly becomes darker as more humus is added. Humus holds water, thus changing the harsh dry habitat and making it possible for other plants to grow, like the Mosses, Liverworts and Lichens. The young unstable white dunes slowly turn grey as they age and are colonised by a continuous mat of vegetation. One of the earliest plants to grow on the surface of the sand is a moss called *Tortula ruraliformis*. Very few mosses have common names simply because most people don't take much notice of them, so you can christen them what you like. My name for this one is 'the all-screwed-up moss' because it is adapted to live on the exposed, dry surface of the sand and spends much of its life all screwed up (fig. 10), but as soon as it rains the leaves unwind, pushing the sand grains out of the way as they uncurl.

Two plants, which can appear very early in the dune succession, are members of the family UMBELIFERAE. This is one of the easiest families to identify with its cartwheel of flowers, each held on the end of a constellation of smaller and smaller umbrella spokes (fig. 11). These two plants are the Wild Carrot and the Sea Holly. The latter tries to disguise its family characteristics, but

Fig. 10. 'The all-screwed-up moss'. On the right, all screwed-up in dry conditions. This helps to reduce water loss.

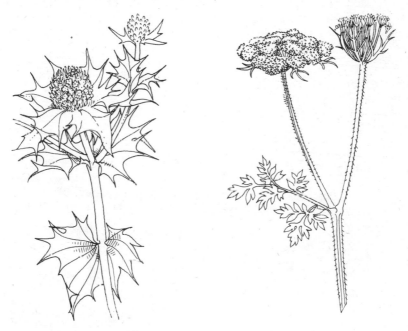

Fig. 11. Sea Holly and Wild Carrot.

does nothing to hide its prickly unpalatability. The Wild Carrot, on the other hand, hides nothing, proud to be a member of the very large Umbel family, growing on the young dunes where there is abundant calcium in the form of tiny pieces of shell in amongst the sand. The Wild Carrot is a CALCICOLE, that is, it thrives on calcareous soils like those of the young dunes. It is a sweet life, but a short one because the combined effects of the acid products of the Marram and of the rain falling on the sand gradually leach the calcium out of the dunes. So the habitat changes and different plants begin to come in.

Passing back from the open sea, the successive waves of the dune ridges tell the whole story, from the harsh life on the white dunes, through the fixed grey dunes, to the stabilised ridges covered with dense natural forest of Oak, Birch or Beech. If you want to see natural forest on sand dunes, you will have to look a bit further than the British Isles, but not all that far because such forests are widespread on the coast of Holland. Holland, a country which used to be very much smaller and still would be except for the plant *Ammophila arenarea*, which spends its time building these castles from the sand of the Kingdom of Canute, and so pushing back the sea (plate 13).

Plate 11. The Kingdom of Canute. That's Holland that is.

Plate 12. Part of a parabolic dune covered in vegetation.

Plate 13. A forest built on sand.

Plate 14. The Grass of Parnassus.

There is, however, much more to the story of the dunes. Each
ridge has its surprises and so do the depressions between the
ridges. It is in the deeper depressions that the rainwater percolat-
ing through the sand collects to form extensive wet areas, and
even ponds and lakes. These wet depessions in amongst the young
dunes, or dune slacks as they are called, are oases in the desert-
scape of sand. The water, charged with the lime washed out of the
sand, gives ideal conditions for the growth of marsh plants. The
slacks often abound with orchids and many other delicate plants
like the beautiful white Grass of Parnassus (plate 14). Actually this
isn't a grass at all, it has a family all to itself called, as you might
expect, the PARNASSIACEAE.

Between the older dune ridges, where all the calcium has long since disappeared, the dune slacks are full of acid-loving bog plants. Heather, heaths, and even the bog moss Sphagnum, all of which loath and detest too much calcium, grow in profusion. The uniform world of sand is not so uniform, for the various faces of the ridges have their own characteristics. The cooler, damp, north-facing slopes abound in mosses and a profusion of herbs, while the warmer, drier, south-facing slopes are dominated by lichens and tiny annual plants which complete their development before the hot dry days of summer.

These same hot days of summer bring the sun-worshippers to the shelter of those south-facing slopes. It is these same dry slopes that are most vulnerable to damage. Anything which disturbs the balance of the living skin of the dunes can initiate erosion. Once the sand is out of the grip of the vegetation, it comes once more under the influence of the wind. The wind is both the alpha and omega, the creater and the destroyer of the dunes. A small patch of open sand produced by a careless boot or picnic fire is a wound in the living system, a wound worked on by the wind like a disease, desiccating, tearing open an ever larger gash in the living skin. Great blowouts develop and then the dunes are on the move, changing shape, slowly migrating across the landscape.

In 1694 the dunes of Culbin, in Morayshire, were on the move; exactly what triggered the process is not really known, but the result was catastrophic. The shifting sands engulfed both productive farmland and a great estate.

The march of the dunes was only finally halted in 1921 by a massive effort of the Forestry Commission which included both the replanting of Marram and the thatching of the dunes with brushwood and finally the planting of a forest.

The great fortresses of sand which give protection to our coasts are not impregnable, so think next time you enter the living world of the dunes and pay at least a little homage to the king in the Kingdom of Canute, *Ammophila arenaria*.

5 *You can't see the Wood*

Next time you sit under a tree and an apple falls on your head, don't just curse, follow the example of Sir Isaac Newton, think about gravity. Thanks to him we know what made the apple fall, but what held it up there in the first place against the force of gravity? A big apple can weigh 4 ounces, so a big apple tree, fruit, leaves, branches, twigs and all weighs over half a ton, and it is all held up by the trunk which is made of wood. Once evolution had taken plants from the limitation of the sea on to the dry land, the struggle for survival became a struggle up towards the sunlight. The problems were many but the potential to the winner was enormous and so gradually the tree habit evolved. Plants developed strong woody stems which could stand upright in the non-buoyant air, not only holding up the leaves to the sun but, in addition, keeping them adequately supplied with water. The tissue which evolved to fit the bill was the xylem or wood, consisting of lots of little tubes of varying diameter joined end to end. They form a network of pipes which run from the roots to the tips of the highest leaves, a distance which may be well over 60 metres. The pipes are highly modified cells and, like all plant cells, they are in fact cellulose boxes. The main feature of these long cells is that they are empty (they must be in order to be able to carry the water) and the cell walls are impregnated with a chemical substance, lignin, which is both a water-proofing agent and a preservative. The cells are therefore dead and they die in the course of duty; their duty is water transport.

The two main components of this mini plumbing system are shown in the diagram (fig. 12). The vessels are the fatter ones; they have definite holes in their end walls and therefore present no

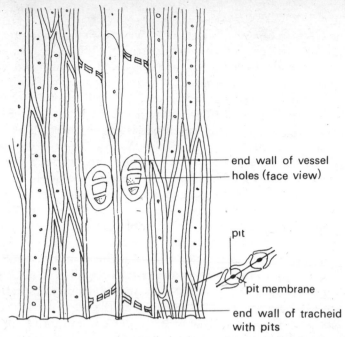

end wall of vessel
holes (face view)

pit

pit membrane

end wall of tracheid
with pits

Fig. 12. Detail of a tiny piece of the inside of a tree tunk. The two main types of plumbing fitments are, the wide ones with holes in the end walls, called vessels, and the narrow ones with pores in the end walls, called tracheides.

bottlenecks to the flow of water. These are the height of evolutionary plumbing, first class lignin-plated fittings. The narrower ones are called tracheides, their end walls don't have holes, only pores which, as you can see in the diagram, force the water to take a more tortuous route, making it squeeze through the pit membrane with its minute holes. Of course, water does not have to squeeze; a water molecule is almost infinitesimal compared to the pit. In actual fact, it would be very difficult for water to 'squeeze', as the molecules are packed together very tightly – but more of that later.

The real problem is not how the water gets through pipes but how it gets up a tall tree. Where is the pump, and if there is a pump, it must be very efficient to be able to lift water in some cases over 200 feet. Remember, anything over 32 feet is going to take some effort because at such a height the weight of water is equivalent to atmospheric pressure, and a simple suction pumping action is not enough.

The German botanist, Strasburger, was the first man to investigate the problem in a big way. His experimental set-up was simple; fell a 75-year-old tree, keeping the saw cut fully saturated with water, then stick the cut end into a bath of picric acid. Picric acid is a very strong oxidising agent and will kill anything. Strasburger's aim was to kill all the living tissues within the tree and see if water transport still carried on. It did. Up went the picric acid and it continued to do so for seven days. (See the results in the table.) *Quod erat demonstrandum*, water transport is a passive process and so a small but important piece of the dogma of the science of botany was created and was welded into all basic textbooks. The passive mechanism depends on the water molecules which are tightly bound to their neighbours by forces – called cohesion. All the time there is a continuous column of water molecules from the reservoir of soil water through the plumbing of the wood up to the leaves, water transport will go gaily on in its own sweet way. Water is lost from the leaves by evaporation, breaking the forces of cohesion only at the top of the water column, so the next molecule in the water column moves up to replace it, pulling all the other ones along behind.

As far as we could discover, everyone quoted from the literature, but no-one had ever repeated the experiment. So we decided to repeat the whole thing, and with the help of a BBC camera crew, the big operator botanists swung into action under the direction of plant physiologist, Geoff Banbury (plates 15–18). An oak tree of the right dimensions was located (by kind permission of The Forestry Commission) in Thetford Chase, and it was swathed in scaffolding. The trunk was severed and throughout the process the cut end was kept saturated with water via a hose pipe. Only those who like getting soaked will really understand what great fun it was. A thick slice of the trunk was then removed from the bottom and the tub was slid into place and filled with the dreaded picric acid; and just seventy-nine years after Strasburger's tree died in the cause of science, so did ours, all 2 tons of it, although of course the wood was dead already. The race was now on and the team was hard put to it to keep up the supply of picric acid. The table opposite shows that our tree won hands down or leaves up when it came to shifting litres of picric acid.

Up went the picric acid; it even began to crystallise out on the dying leaves. The picric acid was then replaced by dyestuffs and

| Day | Volume of picric acid per day in litres | |
	Strasburger's tree	Our tree
1	11·5	41
2	11·5	96
3	9·9	36
4	10·7	54
5	9·8	12
6	6·7	8
7	4·8	30
8	—	15
9	—	8

when the tree was finally laid to rest and sawn up into pieces, the dye was found throughout the tree all the way up to the top, the youngest outer wood being stained the deepest red. Strasburger was right – water transport is a passive process dependent on a dead plumbing system filled with highly cohesive water molecules which supply the process of evaporation from the leaves.

Is that the whole story? Look at the results. In both sets there is an overall decline in rate throughout the experiment. This could be due to very wet still days but in our case it wasn't and we have got the meteorological data to prove it. Another obvious explanation is that as the leaves die, the surface from which water evaporates gradually decreases. The only complicating fact is that the young wood had living cells at the start of the experiment. Is this the explanation for the initial high rates? The fact is that we do not know and would need to do a lot more experiments in order to find out.

One fact does remain, and that is that an oak tree like ours uses an awful lot of water. Multiply it up by the number of trees in a forest and, especially in a dry region like East Anglia, you will see that the rainfall would not go all that far in maintaining an adequate supply. The complex fact is that water loss from the tree can be controlled by the stomatal pores in the leaf which let the water out (fig. 13). However, if a tree has got to keep its stomata shut to keep the water in, then carbon dioxide will not be able to diffuse into the leaf so efficiently and the process of photosynthesis must suffer. Remember that apart from the picric acid, our tree had never had it so good as regards water supply and

Plates 15–18. Re-enactment 1971 of Strasburger's classic 1892 acid bath experiment. The scientist in charge of the experiment was Geoffrey Banbury (in glasses).

cuticle (water proof)
guard cell

Fig. 13. Stomata are pores evolved to let carbon dioxide into the plant, and to control the loss of water vapour. The pore is opened and closed by differential movements of the guard cells.

SO control of water loss was not necessary. Nevertheless, on a hot, dry windy day the rate of loss of water from any forest is enormous and it is easy to see that an extensive forest not only modifies the microclimate (see chapter 7) within the forest, but also can affect the climate of whole regions.

It was probably under the stress of not enough water that plants evolved the – at first sight – stupid habit of chucking all their leaves away. During winter, when the ground and hence part of the water supply is frozen solid, and the winter winds are blowing, a tree with great flat leaves would lose an awful lot of water. The conifers, most of which are evergreen, have much reduced needle-like leaves which are modified against excess water loss. They can therefore ride the storms of winter and keep photosynthesising. That is one reason why the conifers grow so fast, but in doing so, they do lose an awful lot of water. It has been stated, in fact, that if the whole of southern England was planted with pine, the Thames would run dry. A nice story and if our oak tree is anything to go by, there is probably some truth in it.

Wood is fascinating stuff whichever way you look at it. It is interesting that in the days of plastic, the patterns that the manufacturers copy are those fabulous patterns of real wood. The next best thing to a real mahogany table is a plastic replica of a mahogany table. The beautiful pattern of the real stuff is the evolved pattern of all those little pipes produced by evolution to hold up the leaves and keep them well supplied with water.

By the way, conifers only have the narrow, less efficient tracheides and I was just thinking 'are they really less efficient?' You don't happen to have a 60-year-old pine tree you don't want?

6 Decay and Delicacy

It's all a matter of, 'is red cabbage green grocery?' Are mushrooms really plants? All self-respecting plants have the green pigment chlorophyll which is the key to the process of photosynthesis. Chlorophyll traps the light energy which by a complex chemical process is transformed into energy-rich sugar. However, the fungi are undoubtedly plants, they don't rush about like animals (although some of them can move), and on the whole they don't actively eat things (although, again, some of them do engulf their food). Perhaps self-respect doesn't come into it; they do their job and they do their job well. In fact, without them, life on earth would be a pretty messy business, because the fungi (a term which includes the toadstools, mushrooms, moulds, slime moulds and yeasts) are the refuse disposal operatives of nature. They don't need chlorophyll because they have kicked the habit of photosynthesis and live on the productivity of other plants and animals. The fungi are either saprophytes, that is they live on the energy of dead and decaying organisms, or they are parasites, feeding directly on the energy stores of other living things.

Fungi exist in all kinds of places as Alexander Fleming discovered to the benefit of all mankind. Professor Fleming simply left a culture of bacteria lying around and by chance a spore of a fungus *Penicillium* got in. This fungus was well known as a bit of a nuisance because it grew on things where it was not wanted, discarded food, leather, etc. But its importance was that while it grew, it produced a waste product which killed off bacteria. Fleming found the mould *Penicillium* growing within his culture and noticed that round the edge of the fungus colony the bacterial colonies were not growing. On that simple observation hinges the whole science of antibiotics, a science which has saved a great deal of suffering; all because a chance spore floated into the right person's laboratory.

In the same way, the wrong fungal spores only too often get into the wrong place; or if you look at it from the point of view of the fungus, get into the right place. The worst and most spectacular example is dry rot, a common disease of old houses; the fungus, eating into the wooden superstructure, sapping the energy from the timber, can bring about total collapse (plate 19).

Of course, the bulk of the fungi don't bother man at all. They just get on with their jobs down in the woods and meadows, helping to clean up the remains of anything that has remains. The fungus unobtrusively works away, out of sight, deep in the leaf litter or the old tree stumps until it produces a fruiting body which gives the whole show away. Next time you see a toadstool,

Plate 19. Dry rot.

or a bracket fungus, remember two things. First, you are only seeing a small part of the fungus, the bulk of it consists of a delicate weft of filaments (*hyphae*) permeating through the substrate. Secondly, the small part you do see is highly specialised for the one process of producing and dispersing those all-important spores.

The toadstools certainly include some of the most bizarre forms of plant life to be found on earth. Why toadstool? Gnomes, yes, but surely not toads. The name is actually derived from the German 'Tod Stuhl', meaning 'the stool of death'. Some of the toadstools are extremely poisonous, the most deadly being *Amanita phalloides*, the death cap (plate 20).

So if you are going to be a gourmet mycophagist (someone who delights in eating toadstools), and you want to live to a ripe old age, then you must be a good field mycologist (someone who

Plate 20. A Stool of Death, *Amanita phalloides*.

really knows his toadstools). If you are, then a whole new world of exotic tastes is open to you. If your palate craves the taste of mushrooms (*Psalliota campestris*), then beware of becoming hooked on chanterelles, morels and especially on truffles. They can take a lot of finding, especially the latter, which grow underground in the soil of beechwoods in limestone country. There's only one thing for a main-line truffler, and that is a well trained truffle hound whose fastidious nose can lead you to this, the greatest delicacy.

However, it's not just the big fungi which produce the gourmet's delight. *Penicillium roquefortii* is the active ingredient of Roquefort cheese and that blue colour is just the great masses of its spores – what a plant! Best of all is the one that is often entirely forgotten. Yeast. Yeast is nothing more than a microscopic fungus, a single cell which when placed in its correct habitat, buds and buds producing new yeast cells (fig. 14). The yeast plant lives on the energy of sugars and other organic compounds and is able to live in more or less anaerobic conditions. The energy is used in growth and reproduction, and as a by-product of its life processes, two waste substances are produced. One is alcohol, the other carbon dioxide. Alcohol, the basis of a great chemical industry and the great hangover: alcohol which has on both counts made and lost vast fortunes, a waste product of the yeast plant, a raw material of many multi-million dollar industries, brewing, distilling, wine-making, chemicals, etc., etc. All these are possible because this plant can live in a habitat with little or no oxygen and a habitat which contains an increasing amount of alcohol. But even the yeast plant can have too much of a good thing, and as the level of

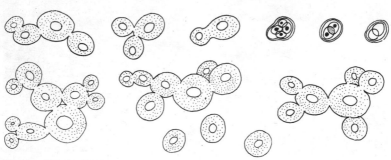

Fig. 14. Yeast cells.

alcohol in the brewer's vat increases, the yeast will die. What a way to go!

It's an interesting point, especially for all teetotallers to remember, that yeast is used in the manufacture of bread. Yeast is kneaded in and starts to grow and reproduce, producing alcohol and carbon dioxide. The carbon dioxide makes the dough rise, while the alcohol – well, of course, it goes up the bakehouse chimney. Man cannot live on bread alone, but with bread and fungi life should not be too bad.

Meanwhile, back in the wood, what about all those toadstools each doing their own thing disseminating the spores of the next generation? Next time you see a toadstool, collect it carefully, break off the stalk and lay the cap right side up on a piece of paper. Leave it overnight and carefully remove the cap in the morning. The result may be a beautiful spore print. Two precautions are necessary. It may have white spores, in which case black paper is best, and don't leave it in the reach of children just in case you happen to have collected the wrong one. Now try and count the spores. The number takes a lot of believing (one giant puff-ball produces a mere three trillion), but the actual mechanisms of dispersal are even more unbelievable.

Take, for example, the common field mushroom. Lining the gills are batteries of little spore guns (basidia), each loaded with four spores (fig. 15). When ripe these are shot off with 'great' force, great, that is, relative to the size of the spore. Now if they are shot off too forcibly, then they will hit the adjacent gill and be trapped. So the range of the guns is very accurate, and the spores are only shot to half the distance between the gills. At

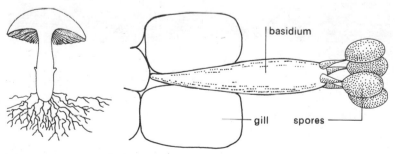

Fig. 15. A battery of Basidiospores.

Plate 21. A living bracket.

the end of their trajectory, they fall vertically under the pull of gravity. This means that the gills must be absolutely vertical; if not the whole spore system will soon be gummed up solid. So the whole life of the above-ground bit of the 'toadstool' is orientated by gravity. Just how, we are not sure, but we can prove it by growing a mushroom on a clinostat, which is an instrument which simply turns the developing mushroom round and round so that it is equally exposed to the force of gravity from all directions. Then things really start to go wrong, and believe it or not, the gills can be formed on the top of the umbrella.

Now this will be even more critical in a thing like a bracket fungus with pores rather than gills, and this will be especially so if it is perennial. This means that the formation of the pores must all the time be orientated with respect to gravity and this always happens. Next time you find a decaying tree with bracket fungi growing out of it, look round for a fallen branch and you will find that the fruiting body, which was perfectly orientated to gravity on the standing tree, has changed direction. A whole new bracket has begun to grow reorientated in its new location to, again, take advantage of the effect of gravity for dissemination of its spores (plate 21).

The role of a toadstool is to produce spores, get them up, shoot them out and drop them into the moving layer of air above the soil surface. Then there is a chance that they will be carried to a new habitat of the right sort. The whole evolution of the fungi, or at least of its fruiting body, centres around dispersal of these all important spores, and evolution has produced some 'way out' things.

The puff-balls consist of nothing more or less than thin parchment-like bags of spores (fig. 16). As the spores and the fruit body ripen, the top of the bag disintegrates leaving a small hole. The wind blowing on the outside of the bag, or better still, drops of rain falling on the bag make it act like a pair of bellows, puffing out the spores. If you want to, you can try for yourself, because puff-balls, especially *Lycoperdon esculentum*, are not uncommon even on the edge of garden lawns.

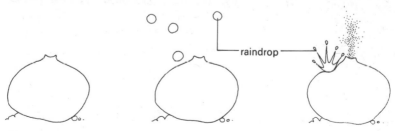

Fig. 16. How a puff-ball puffs.

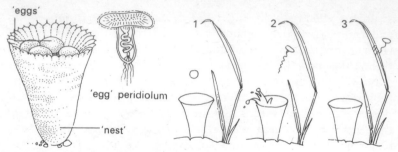

Fig. 17. A Bird's Nest fungus.

Rain splash is also the mechanism upon which spore dispersal in the Bird's Nest fungi depends (fig. 17). Here the spores are in the form of little eggs, each with a spiralled tail; they are called peridiola. The peridiola lie in the nest until they are ripe. Raindrops falling directly in, dislodge the eggs with amazing force, splashing them out to distances in excess of two feet. Here the spiralled tail comes into play, winding round and fixing the spores to grass leaves. These may in turn be eaten by animals, the spores being eventually deposited in the dung a long way away – a complex but efficient dispersal mechanism.

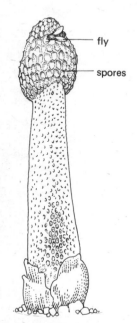

Fig. 18. The Stinkhorn.

Many other fungi depend on animals for their dispersal and they go the whole hog, living on the dung; these are the Coprophilous fungi. The spores of these fungi landing on herbage are eaten and pass unscathed through the alimentary tract of the animal to emerge and live it up on the dung. It may be rather an extreme case of everything to its own taste, but it does remind us that nothing goes to waste in the world of nature.

It would be impossible to talk about the dispersal of fungal spores without mentioning the Stinkhorn. This, aptly named *Phallus impudicus* (fig. 18), is quite common in a whole range of woodlands, and you always know it is there long before you see it because of its smell. To call it pungent would be to put it mildly. The fruiting body emerges with extraordinary rapidity from an innocent looking white egg about the size of a golf ball. The spores, a khaki green sticky mass, are borne at the top of the gleba and the smell attracts not only all the marauding mycologists, but all the flies of the neighbourhood. These crawl all over the gleba (the flies that is, not the mycologists), inadvertently collecting the sticky spores as they go. Soon all the smell and the spores are gone, transported to a new habitat where they have a chance of thriving, if they find a new source of energy of the right kind.

Next time you see a toadstool, don't just kick it over, leave it there to mature and think how its own particular function of spore dispersal has evolved making use of whatever opportunities the environment had to offer.

7 Where have all the trees gone?

The only real problem with being a tree is that you are much too useful. Wood, the original plastic, can be used for everything.

Now we know, mainly from pollen studies (see chapter 3), that only 6,000 years ago the bulk of lowland Europe, including the British Isles, was covered in dense forests of Oak, Ash, Elm, Lime and Hornbeam. But then along came man, changing the face of the landscape by removing the trees. In those parts of Europe which had an equitable climate the effect was not too disastrous, but in parts which suffer (or enjoy) a hot dry summer, the effect was catastrophic. The hot baked landscape of the central Meseta, the Spain beloved by artists and tourists alike, is a prime example, the perfect place to tilt at windmills but a hard task-master for the farmers.

While planning a botanical expedition, we decided to go and look for flowers in Spain, and literally stuck a pin in a map. It landed almost on the highest peak of the Sierra de la Cabrera which is shown on the map (fig. 19) and La Baña, a village at the end of a long dirt road was chosen as base camp. The terrain was rugged, and eight miles away from the village a small lake almost filled the valley head (plate 22). There, nestling beneath steep cliffs on the edge of the lake, was a scrap of forest. At first sight it was not much to look at, but once inside it was found to be a phenomenon in its own right – in fact, a piece of the original forest, the only one that had survived the ravages of man and of time. Inside the forest it was really a Walt Disney kind of world, the twisted tangle of old branches almost as impenetrable to the would-be explorer as the forest canopy was to the sun. In the dense shade were a fantastic range of plants, deep carpets of moss covered the forest floor, splashed by the streams and the sun-flecks. The trunks and branches of the trees were draped with a profusion of lichens which included the large hanging 'Spanish Moss' and the 'Lungwort', showing that atmospheric pollution is here a thing of the future. (None of the large lichens, and especi-ally these two, can tolerate sulphur dioxide which is a main pollutant near towns and cities. The reason is simple. If a lichen lives out its life as an epiphyte on a tree trunk or growing on bare rock, it is out of contact with the nutrient rich soil. In actual fact, lichens must depend in the main on the nutrient salts dissolved in the rain which, under natural conditions, are precious few. So evolution has made them highly efficient at catching any-thing which comes their way, and in a polluted atmosphere that's just too bad.)

Wherever light did penetrate the canopy, not an inch of space was wasted. The ground flora included such common garden plants as London Pride and Solomon's Seal, mixed up with

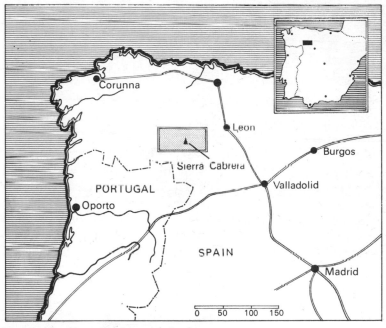

Fig. 19. The Sierra Cabrera and the forest.

Plate 22. The forest refuge. Note the dead Holly trees in the foreground.

peculiarities like Herb Paris, *Paris quadrifolia* (fig. 20). This plant, long held in esteem in witchcraft, has a peculiar yellow-green flower which smells of rotting meat and is pollinated, as you might have guessed, by carrion flies. One British plant is well worth a special mention. That is the Bilberry or Whortleberry, which is a common and dominant plant in our open countryside. However, even high in the mountains of North Spain it was only found in the shade of the trees. All these and a host of other plants were confined to the forest, and this small scrap of woodland represented their last refuge in a vast tract of barren, rocky heath.

Throughout the valley, wherever the stream waters draining from the snow slopes above were ponded back, small peat bogs (see chapter 2) – hanging drops of history – were found. Study of the peat profiles showed that the forest had been much more widespread in earlier times. All the evidence pointed to the fact that the first phase of deforestation of the valley was contemporary with the appearance of man on the scene. The great wooden beams in the houses of La Baña were a good indication of where some of the trees had gone, and it is not impossible that some of the choicest oaks went to sea with the Armada. It is an interesting fact that at least two of the trees still living in the relic forest predated the Armada by over 200 years, one of these was a Birch (girth 5·9 metres), the other a Yew (girth 6·2 metres).

Why had this tiny piece of forest been spared? Well the answer in all probability lies partly in its structure. The top storey or canopy trees are Birch – these are the sun plants which have to bear the full blast of the summer sun as well as the impact of the icy winds of winter. Below these is an understorey of Willow, Holly and Yew. The latter is undoubtedly the most abundant and probably holds the key. The leaves of the Yew are poisonous, not only to us but also to farm stock. So the local shepherds have a problem. If the Birch trees are cut for fuel or for fodder, then the

Fig. 20. Herb Paris with detail of flower.

flocks will have easy access to the deadly Yew, so the best thing to do is leave the lot alone. Whether this theory holds water or not, the fact remains that this is the only bit of forest left.

The pattern of deforestation of the valley, and indeed of many of the slopes of the Sierra de la Cabrera, must have followed something like the following pattern. Man came into the area and lived as a hunter, reaping the benefits of the productive forest landscape. As soon as he had learned enough about the potential of the area, he started to clear the forest for primitive agriculture, growing crops on the better soils and grazing his animals on the steeper slopes. The forest was still of great importance, timber for building, a ready supply of food, and leaves to be collected as winter fodder for the animals. (There is much evidence that early Neolithic man in Britain used this method to supply his cattle with winter food; in fact in Britain this was practised down to the seventeenth century. The fascinating thing is that it is still going on today in La Baña, and you can see all round the village the stores of leaves and leafy branches.) Gradually the highly productive forest was pushed back and replaced by a plant community dominated by the Tree Heather, *Erica arborea*. As the process continued, the villagers had to graze their cattle further and further afield, and the shepherds had to start burning the heathy vegetation to encourage the growth of new young shoots on which the animals could feed. So it is today, with the shepherds taking the goats and sheep, which belong to the whole village, out to the pastures, often travelling thirty miles in one day in search of sufficient food for their flocks. Today this tiny village draws on the resources of an enormous area of countryside which is rapidly becoming more and more like a desert. Burning and grazing, burning and grazing, the cycle is as relentless as the hunger of the goats. The Tree Heather can take a lot but the end must come and then not even the woody rootstock is allowed to go to waste, but is grubbed up and used for winter fuel (plate 23). Nature does her best to repair the damage; the only plants which survive are those which are inedible and hence not worth burning – plants like *Genista hystrix*, which looks like, feels like and in all probability tastes like barbed wire.

The pressure is really on and even the tiny scrap of forest is now being attacked. The canopy Birches of the forest edge are being cut and their leaves used as fodder, the stark tale of destruction

Plate 23. Woody rootstocks of the Tree Heather stored for winter fuel.

being adequately documented by the bare white skeletons of the
Holly trees which die as soon as they are exposed to the full impact of the sun.

It was these trees which made us ask one question and gave us a
way of answering another. The Holly trees can tell the difference
between the environment inside and outside the forest; question
one, exactly what is the difference? The effect of man has been
to replace a productive forest with unproductive heath communities; question two, just what is the difference in production?
Well, with dead trees, recently felled trees, old tree stumps and a
lot of expert and willing help from the members of our expedition,
it was possible to get at least a rough answer.

The forest trees were identified, counted and measured, old stumps were cleaned, measured and aged by counting their annual rings. Hence every tree in the forest could be approximately aged. Dead and recently felled trees were dismembered and weighed and a rough age/weight relationship for every species was built up. In this way, figures for standing crop and net annual production of the forest were determined. Corresponding figures for the heathy replacement communities were obtained in the same way. Throughout the whole expedition crude but effective meteorological stations were maintained within each

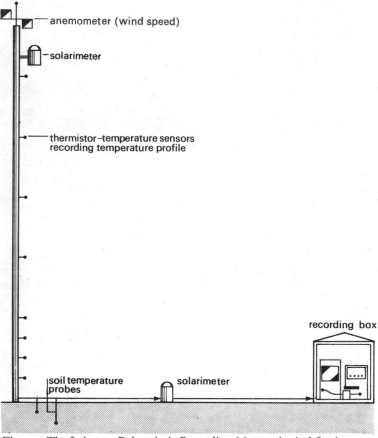

anemometer (wind speed)

solarimeter

thermistor–temperature sensors
recording temperature profile

recording box

soil temperature probes

solarimeter

Fig. 21. The Leicester Polytechnic Recording Meteorological Station.

main type of vegetation. The data from these were supplemented during a later visit by recording stations (fig. 21).

The results are indicated in figs. 22 and 23, and can be summarised as 'fantastic'. The effect of man has been to replace a forest with a potential production of 3,500 grammes in one square metre each year (1,600 leaves, 1,900 wood) by heath communities, the productivity of which ranges from 200 to 300 gms/m²/annum.

Study also showed that in the less easily accessible areas round the forest natural regeneration was taking place, the ground being

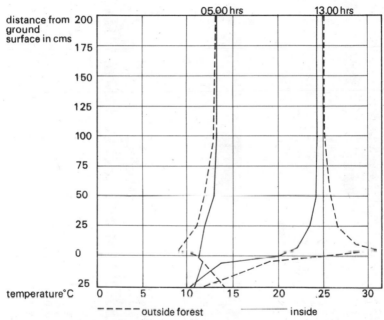

Fig. 22. Temperature profiles inside and outside the forest recorded at 0500 and 1300 hrs on 8 June 1971.

prepared by the growth of the Broom (*Sarothammus striatus*), which has nitrogen fixing bacteria in its roots. Once the Broom was re-established, the forest trees were not long in following. So given time and if left undisturbed the forest could in all probability re-assert itself, producing once again a productive forest landscape with a much more equitable climate beneath the forest canopy.

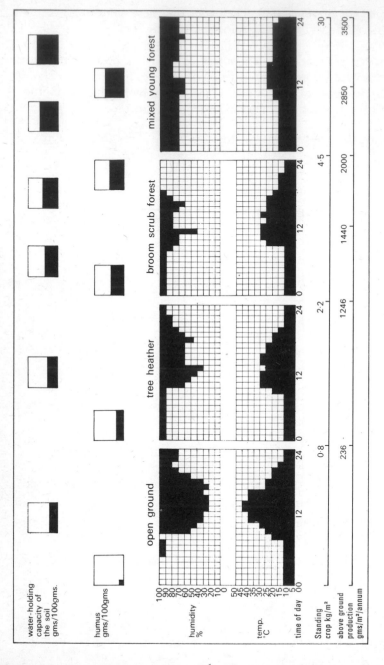

water-holding capacity of the soil gms/100gms.

humus gms/100gms

mixed young forest

broom scrub forest

tree heather

open ground

humidity %
100
90
80
70
60
50
40
30
20
10

temp. °C
50
45
40
35
30
25
20
15
10

time of day
00   12   24   0   12   24   0   12   24   0   12   24

Standing crop kg/m²
0·8   236   2·2   1246   4·5   2000   30   3500

above ground production gms/m²/annum
236   1440   2850

What will happen? There are at least three possibilities. First, that things will not change and gradually the area will become a rock desert of no use to man or nature. Secondly, that the area will be re-forested, in all probability with exotic conifer species. Thirdly, that conservational management aimed at replacing the natural forest cover could be introduced. Depopulation is not the answer, it is only an admission of defeat. If the community of La Baña disappears, with it will go all the local expertise and knowledge which has evolved with its people in relation to that environment.

The real problem is how to put the trees back without putting the clocks back. The only real answer is diversification of those very skills within the area so that absolute dependency on the destructive flocks and land use practice is reduced, thus allowing a gradual reinstatement of the forest.

Don't run away with the idea that this is a problem unique to La Baña or to Spain. The only difference is that in other parts of the world where the climate is less harsh, the effect is less rapid and less catastrophic. The great palls of smoke which often hang over the Pennines and vast tracks of the Scottish Highlands tell the same tale. The burning, grazing cycle, whether it is for aristocratic grouse or subsidised sheep, must in the long term convert our uplands into wet deserts. Yes, even in Britain, the process of depopulation of our uplands goes on, admitting defeat in this struggle between man and nature.

Fig. 23. Left, data concerning the probable course of regeneration showing how the build-up of standing crop (trees, etc.) and humus, modifies the soil and the climate, leading to a productive mixed forest. Climatic data recorded one hourly intervals over the hottest day of the expedition using wet and dry bulb thermometers fixed 10 cms above ground.

8 *In the Swim*

Life began in the sea and among the first living organisms were green plants which could capture and use the energy of the sunlight. However, a lot of energy falls on dry land and it is not surprising that part of the main drive of evolution took the plants from the sea to exploit the potential of the land.

To do this plants had to evolve thick water-proof skins to prevent excessive water loss, pores to let carbon dioxide get in (but these pores, called stomata, must also regulate loss of water vapour), a vascular system to keep the shoot supplied with water, a root system for anchorage and to tap the soil for water, and strengthening tissue to hold the plant up in the non-buoyant air. These features allowed the plant kingdom to take its major step out of the salty sea on to the dry land and the most successful group of land plants are the ones with flowers, the ANGIOSPERMS.

Then a funny thing happened; some of the land plants returned to water, fresh water. What prevented the seaweeds from making the direct switch, we don't know, and very few of them managed it. However, the ferns and the flowering plants took up the challenge of life in sweet water. This meant a real shake-up in evolution, all those major features which fitted the plants to life on land became redundant; back in water life does not depend on cuticles, stomata, strengthening or conducting tissue nor on a root system. Water loss is a non-problem, and absorption of nutrients can take place over the whole surface of the plant. There are, however, two new problems – the waters and muds of small lakes and ponds can become depleted of oxygen, and flowers need pollination, a process which is rather difficult under water where there is no wind and no bees.

Evolution solved these problems, remodifying the form and structure of the land plants to produce the successful hydromorphs (water plants). Hydromorphs come in all shapes and sizes, but they all have one feature in common, they all have aerenchyma (fig. 24).

Aerenchyma is a tissue of holes held together by living cells in the same way that a knitted thermal vest is a lot of holes held together by wool. As a mechanism for flotation and a system for storage and transport of oxygen-rich air around the plant, the

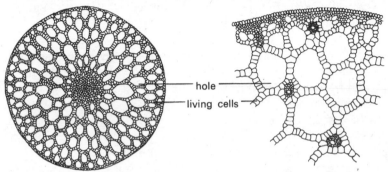

Fig. 24. Nature's own, 'geodesic air frame' aerenchyma, holes held together with living cells, maximum strength with minimum use of resource.

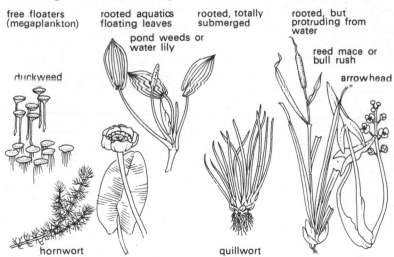

Fig. 25. Evolved to be in the swim. Water plants show a fantastic range of structure.

aerenchyma appears to be perfect. The true functional signifi-cance is, however, more subtle. In fact, it is not such a perfect transport system because the rate of gas transport will be limited by the size of the smallest hole, and the air spaces are separated by sheets of cells which have only tiny holes between them. The buoyancy and strength of the water plant is attained by the geodesic structure of the aerenchyma and, as in an aircraft structure, this gives maximum strength using the minimum of material. That is the key factor; living cells need oxygen so the fewer cells used, the less oxygen required – it's just what the evolutionary doctor ordered.

FREE FLOATING PLANTS (Megaplankton) only have one major problem – they are at the mercy of the elements and can be carried by wind or current to the wrong place, ending up on dry land, or in the sea.

The best examples of surface floaters are the Duckweeds (fig. 26) which include the smallest of all the flowering plants, *Wolffia arhiza*, less than a millimetre long. It's nothing more than a pad of green aerenchyma with no roots. Pollination is no problem as the tiny flower is borne above the water surface. It's very difficult to see a duckweed's flower but in some places in Britain our com-monest Duckweed, *Lemna minor*, which grows in open ditches, produces flowers which are well worth looking out for in June or July.

A Duckweed which floats just below the surface is the ivy-leaved species, *Lemna trisulca*. The truly pelagic magaplankton face the problem of underwater pollination. The Hornwort *Ceratophyllum*

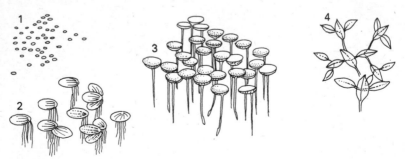

Fig. 26. Variations on the Duckweed theme: *1* Wolffia; *2* Common; *3* Gibbous; *4* Ivy-Leaved.

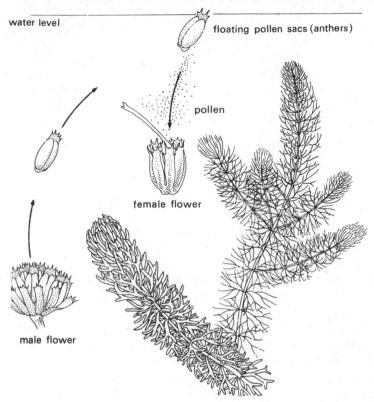

water level

floating pollen sacs (anthers)

pollen

female flower

male flower

Fig. 27. The Hornwort, one of the few flowering plants able to carry out pollination under water.

demersum (fig. 27) bears its small flowers underwater. Once the flower opens, the anthers detach and float to the surface, where they dehisce and shed their pollen. The pollen is heavier than water and sinks back to pollinate the flowers!

Because of lack of light, plants which are wholly rooted and have neither floating leaves nor tall leafy shoots to protrude above the surface, are confined to shallow water. The Water Soldier (*Stratiotes aloides*), gets the best of both worlds. It spends the winter sitting on the bottom, weighed down by deposits of chalk in its tissues. In spring, the plant unships its chalky ballast and rises to the surface where it produces white evil-smelling flowers (fig. 28) which are pollinated by flies.

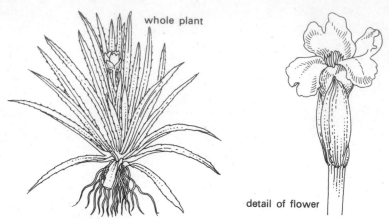

whole plant

detail of flower

Fig. 28. The Water Soldier.

ROOTED PLANTS WITH FLOATING LEAVES

The best example is found in Britain only in hot houses at botanic gardens – *Victoria amazonica*, the giant waterlily of the Amazon Basin.

This enormous plant appears to be an annual, the seed germinating to produce a massive rooting system which anchors the plant firmly to the bottom. Long pedicles extend to the water surface where the leaves expand with amazing rapidity; in a matter of four days a leaf can be as much as two metres across. Apart from its size, the most striking (I use the word with feeling) feature of the plant is the fact that the leaves, leaf stalks and buds are covered with short stiff, highly efficient spines. While filming it in the waterlily tank at the Edinburgh Royal Botanic Garden, we saw the necessity for the armour. If we floated any other plant in the tank it was rapidly devoured by a ravenous horde of herbivorous fish that live in harmony with the queen of the waterlilies.

Even the flower is protected by spines throughout its short life. The spiny flower bud opens to reveal an enormous mass of pink petals, the pink rapidly fades to white and the flower closes for good, pollination having been effected. The whole process usually takes less than one day.

Two factors aid pollination by helping to attract insects; one is a rather revolting fruity aroma, the other is the fact that the temperature inside the flower is higher than that outside. The

Plates 24 and 25. The underside of the leaf of the giant waterlily, showing the massive buoyant ribs.

temperature difference we recorded at Edinburgh was 4·20°C.

So there would appear to be no pollination problems for *Victoria amazonica*, but while making the film we decided to investigate one very obvious problem the world's largest lily has to face. The leaf is not a flat pad, it is bowl-shaped. So why doesn't the leaf fill with water when it rains, and how does it empty? With the help of Dr Brinsley Burbidge of the Botanic Garden we did a little research.

The great leaf is perforated by numerous tiny holes and we found that if we filled the leaf with water and covered it with a polythene sheet to stop evaporation, the water rapidly drained away. Why then doesn't the weight of the leaf resting on the water surface force water up through the holes? (plates 24 and 25). The answer was, of course, simple as most evolutionary answers are. The veins of the leaf run inside great supporting ribs which, being well supplied with aerenchyma, float the actual lamina clear of the water surface, thus allowing free drainage through the pores. Proof came by simply weighting the leaf just sufficiently to depress the lamina below the surface and small fountains of water issuing through each pore soon filled the leaf.

THE STICKER-OUTERS

Many water plants face one great problem which is that once their leaves have filled all the available water surface, their further growth and development is greatly curtailed.

Evolution has overcome this problem by producing the emersed hydromorphs – plants which can grow out above the surface of the water.

The Marestail, *Hippuris vulgaris* (fig. 29), is a good example of a sticker-outer which exploits both environments. Rooted in the bottom muds, its stem, which is well supplied with aerenchyma, bears beneath the surface great masses of long, thin, translucent leaves with neither cuticle nor active stomata. Above the surface of the water the leaves are short, thick and stiff, having both active stomata and a thick cuticle. It is in the axils of these upper green leaves that the small green flowers are found. Incidentally, these flowers prove that this is the Marestail and not a Horsetail (*Equisetum sp*). The Horsetails which to the uninitiated look very much like *Hippuris* are, in fact, relatives of the ferns, therefore having no flowers.

So adaptation through evolution has produced the successful hydromorphs to exploit the potential of the fresh water habitat. Of all the stories of evolution, the hydromorph highlights the real meaning of the two concepts, potential and success.

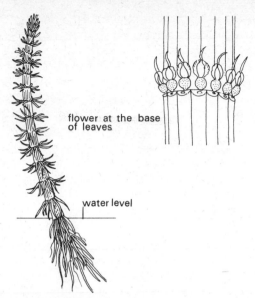

flower at the base
of leaves

water level

Fig. 29. Marestail, *Hippuris vulgaris*.

A VICTORIAN CAUTIONARY TALE

In 1842 a delicate hydromorph by the name of *Elodea canadensis* was introduced to England from, yes, Canada. The enormous potential of our ponds, rivers and especially our then busy canal system was open to exploitation. The new immigrant, in true pioneering spirit, took up the challenge, seized the potential and went west, east, north and south wreaking havoc by blocking up Britain's waterways which were the lifeblood of our developing industries. Share prices were in danger and then, horror of horrors, the blood of every true Victorian gentleman froze, the Varsity Boat Race was in danger, practice on the Cam was impossible.

The threat of the Canadian pond weed did not last long, it died away almost as quickly as it had come. However, the exact reason is obscure. Perhaps the strict Victorian society stifled the North American visitor or perhaps it died of another form of frustration as it appears that only the female plant ever got to Britain.

Who knows? But whatever the reason, the story of *Elodea* is a story of success.

More recently another hydromorph *Eichornia crassipes* has taken over the role of *Elodea* and has been and is causing havoc in tropical and sub-tropical waters. *Eichornia* the Water Hyacinth, a native of Brazil, is a very beautiful plant and is highly cherished by water gardeners, so much so that it is easy to see (fig. 30) why it is so successful. It is both a member of the megaplankton (look at the leaf stalks modified into bulbous floats) and it is emersed, growing up above the water surface.

Just like *Elodea*, wherever *Eichornia* has been introduced into water in which it has no competitors, it has seized the opportunity of unrestricted growth. In so doing, it has blocked rivers and lakes, increased evapotranspiration and hence water loss from reservoirs, provided shade and hence a habitat for mosquito larvae (remember mosquitos carry the dreaded malaria) and has provided food for tens of millions of water snails which can carry a parasite, causing the disease bilharzia. Man does not learn by his mistakes and it is definitely a mistake to play about with such a highly successful product of evolution as a water plant.

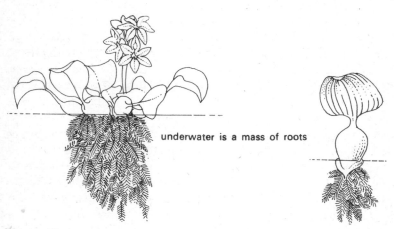

underwater is a mass of roots

Fig. 30. Water Hyacinth.

9 *The Wars of the Primroses*

Primroses are as much a part of Spring as April showers, but how long will it stay like that?

Between 1950 and 1960 (in fact, it's still going on) botanists, amateurs and professionals alike, roamed the face of Britain, each one armed with a sheaf of little white cards. They hunted, they checked and they recorded all the plants which they could find in hedgerows, roadside verges, woods, meadows, in fact every higher plant that grows in every 10km square shown on the ordnance survey map. Not a species was left unturned. This was no mere rabble, it was an army, masterminded by Dr Frank Perring and a high-ranking committee of the Botanical Society of the British Isles. This army of flower people took their 'square bashing' very seriously and the great task was accomplished. Reputations were made and lost as the sacrosanct county and vice-county boundaries, which had till then formed the basis of all plant records, were transgressed by the computerised 10km grid.

All this effort was not in vain. It produced the data for a unique book, *The Atlas of the British Flora*. The distribution of each species is shown separately mapped as a scatter of dots, each one placed in the centre of a 10km square, thus tantalising anyone who doesn't know exactly where it grows.

Remember some of our plants only grow in one place in the whole of Britain, and in these cases the dot in one square means a single record. Others are much more widespread and occur in almost every square, and each dot represents many records. The point is illustrated by the two maps (figs. 31 and 32) which show the distribution of two members of the Primrose family (the PRIMULACEAE), the common Primrose (*Primula vulgaris*) and the

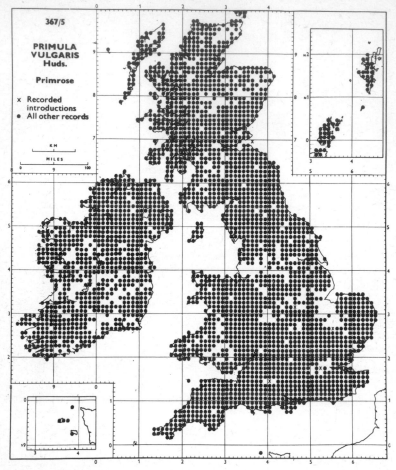

Fig. 31. Distribution of the common Primrose.

not-so-common Oxlip (*Primula elatior*). Yes, if you thought you had Oxlips growing in your favourite bit of countryside and it isn't near the area shown, take another close look and you will probably find that it's a Cowslip or a Cowslip Primrose hybrid you have been looking at. Why is the Oxlip so restricted? They appear to like the damp, moderately well-managed woodland of the clay soils of the Essex/Suffolk borders. Their woodland home is their stronghold, and they need it because their ever-so-common relation appears to be edging in.

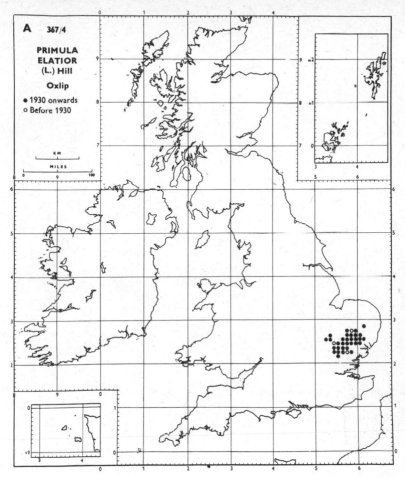

Fig. 32. Distribution of the Oxlip.

How to tell the Oxlip from the Cowslip

	COWSLIP	OXLIP
Fruit	Oval, enclosed within sepals	Oblong, protrudes beyond the sepals
Sepals	Uniform pale green	Midribs darker green
Petals	Folded in the throat	No folds in the throat

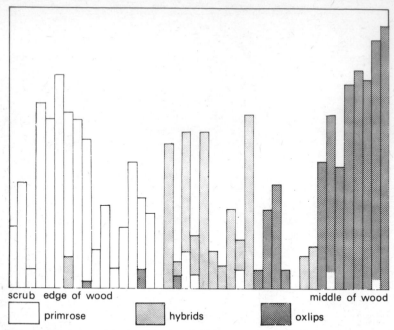

scrub   edge of wood   middle of wood

☐ primrose   ▨ hybrids   ▧ oxlips

Fig. 33. Statistics of the battle ground (Primrose versus Oxlip) at the edge of the wood.

The Primrose and Oxlip are very closely related and inter-breeding is possible wherever populations of the two come within pollinating distance of each other. On the edge of many of the woods it's all happening, and hybrids are found in profusion. Passing into the wood from the open edge, first there are great drifts of real primroses and then the discerning eye, or nose, will pick out hybrids, while deep in the wood only the true blue-blooded Oxlip can be found (fig. 33).

The war of the Primroses was first studied in detail by a local botanist, Miller-Christie, who scoured the countryside reviewing the troops and the state of play. Christie reckoned that the Prim-rose was winning the battle, but a closer look more recently has shown that in some places one appears to be winning, while in other woods, the other one has the upper hand. One reason for the apparent change in the balance of power could be that wood-land management practises, and hence the habitats, are them-selves changing. You see many factors can affect the success of either plant. Take, for example, the simple fact that each seed produced by the Primrose carries a little store of food which attracts ants, and the ants carry them about aiding their dispersal, for instance, carrying them deeper into the wood. On the other hand, the seeds of the Oxlip have no such store of food and depend on the wind for their dispersal, and remember there's not likely to be much wind deep in the wood. So in the quiet of the English countryside, the war of the 'Primslips' and the 'Oxroses' goes on.

There is however, another war of the Primroses which the maps in the atlas do not even show. Nevertheless, its progress has been studied for a very long time by Dr Jack Crosby, a product of the famous Cambridge Botany School. But more of that later, and back to the basic principles of floral structure. Next time you get a bunch of Primroses take a close look. There are two sorts of flowers, pin-eyed and thrum-eyed. The diagram (fig. 34) explains the difference; the reason – well it's all to do with the mechanism of pollination.

The Primrose flower not only attracts humans, it also attracts insects and it is the insects which carry the pollen from one to another, effecting pollination. Imagine an insect with a long proboscis which is poked in search of nectar into the bottom of a thrum-eyed flower. If the anthers are ripe, pollen will collect on

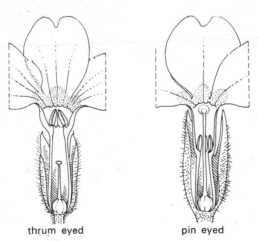

thrum eyed   pin eyed

Fig. 34. Pin-eyed and Thrum-eyed Primroses.

the proboscis, especially near the top, so you see that when it moves to a pin-eyed flower, the stigma of which is receptive, pollen will be very easily transferred to the stigma, ensuring pollination. Of course, it will happen the other way round if it visits a pin-eyed flower first. So cross-pollination between pins and thrums will be the order of the day and the balance of the two types will be maintained in the population, or so Darwin thought, but is it true?

Now down in the depths of Somerset something has happened and a population of homostyles has invaded the scene. Homostyles are simply Primroses with the stigma and the anthers held at the same level, pollination should then be bang on every time and so the homostyle should be at an enormous advantage. If this is true, then homostyles should spread like wildfire.

Dr Crosby has been counting and mapping the populations for years and has found that while the obvious is happening in some places, in others it is not. Why? One possible and ingenious explanation is slugs. If a slug crawls across a Primrose flower in search of food, the succulent anthers will be the main meal ticket and if there is a stigma there it may be eaten as well. A thrum-eyed flower will lose its anthers and a pin-eyed flower will lose nothing at all. The homostyle may, however, lose both stigma and anthers and its flower will then be useless. The balance of the pins, thrums and homostyles may well depend on the balance of the slugs.

You see, there is quite a lot to know even about a common plant like the Primrose. If you have got primroses growing near your home, why not investigate?

1 Are there any homostyles?
2 What sort of insects actually visit the flowers?
3 How long do your primrose plants live and how many flowers do they produce?
4 Are they disappearing because of picking?

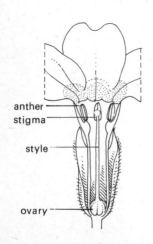

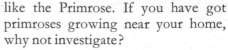
Fig. 35. The homostyle, pattern of the future?

The last one is important because everyone picks Primroses. Look back at the map; is that hole in the distribution pattern developing around London, due to overpicking? If it is, then it could be possible that in another century the map of the Primrose will look like the contemporary map of the Oxlip, and Spring will come and go without the wild Primrose. This is the real war of the Primroses.

10 *What is Vegetation?*

Not twenty-five miles from central London, within the net of sprawling suburbs, amongst the elegant farms of the stock-broker belt stands Box Hill (plate 26). This is just the place to water a poodle or do a spot of courting or just to sit, it is also the perfect place to ponder evolution.

Charles Darwin put forward the theory of evolution through natural selection, a theory which has been contained within the phrase 'survival of the fittest in the struggle for existence'. You

Plate 26. Box Hill, general view.

might therefore expect that each part of the landscape, each habitat, would be filled with one sort of plant, the fittest for that habitat, living it up on the euphoria of evolutionary success. This does not happen, however. Monocultures are rare in nature and most natural habitats are occupied by complex associations of many different sorts of plants; such complexes we call vegetation. What has vegetation to do with evolution? What has evolution to do with vegetation and why bother to think about it on the top of Box Hill?

From the top of Box Hill you can see a large slice of terrain and that terrain is part of the arena of evolution. Most of it screams 'touched by human hand', straight lines and sharp corners, the angled world of man stretches from the horizon almost to your feet, but there at the bottom of the hill the scene changes abruptly. The angles are replaced by blurred rounded masses of vegetation, a continuum of changing colours, textures, heights and shapes merging together. This is not virgin vegetation, it has for many thousands of years been subject to some disturbance and management. Yet, despite a little bit of forestry here and a little bit of park your car and have your picnic there, the natural processes of succession have been allowed to keep going and the result is best called semi-natural.

Look at the range of vegetation which can be found on the hill, an enormous variety in a very small area. It is this very variety which poses the question 'how did this diversity evolve?'

The easiest way to begin to understand is to look at a simple system, one which approaches the Darwinian idea of a single 'super' plant. In the man-made landscape at the foot of the hill, there are many fields each of which contains a well-managed crop, for example, the field of beans. It is planted by the farmer for one purpose – to provide a good crop of nothing but beans.

As a bean plant grows, it produces new leaves at the top and the 'leaf area index' (i.e. the area of leaves per unit area of land surface) gradually increases. This means that there is a greater area of leaves to intercept the sun's energy and hence an increased amount of photosynthesis, thus increasing the growth rate of the crop. The problem is that the leaves at the bottom get into deeper and deeper shade until eventually there is insufficient light for them to maintain an adequate level of photosynthesis. These leaves will now either begin to drain energy which is being fixed by the younger leaves, or they will wither. It is usually a bit of both but the end result is the same; the bottom leaves start to die off. So eventually a stable state will be reached in which leaves are dying at the bottom as rapidly as they are being produced at the top. The leaf area index is then more or less static and the rate of growth of the plants and the crop gradually slows down and finally comes to a standstill. This is just hard luck for the bean plant, and for the farmer – the total production of his monoculture is fixed at that point.

However, there are still plenty of nutrients left in the soil (especially when you realize that the bean plant has nitrogen fixing bacteria in special root nodules), and there is still a lot of spare light even at ground level within the crop. This light is available for photosynthesis provided that there is another sort of plant which has evolved to use it. Even in the perfect farmer's field there are weeds, some of which can live in the gloom of the fully grown crop and others grow either early or late in the year when they are not competing with the beans for the all important light energy. Together the weeds and the crop reap more of the actual potential of the field and this is partly what vegetation is all about.

On Box Hill itself the most complex type of vegetation is the mixed forest found on the summit. Fig. 36 is an example of neo-classic 'lollipopology' but it does give some idea of the complex structure of the forest. The trees – Oak, Ash and Beech are the sun plants which form a patterned canopy, intercepting much of the incident energy. Beneath these are the understorey trees, Yew and Box and, in some places, Juniper, living it up in the cool shade together with the saplings of the canopy trees. Lower still, the shrub layer contains juvenile stages of all those above and a whole variety of other plants, Blackberries, Dogwood, Wayfaring and Spindle tree, Privet and beneath this the rich carpet of herbs. Still lower Mosses and Liverworts can be found in profusion growing close to the rich damp soil; these are the shade plants *par excellence*. Near the floor of the forest two factors may help to compensate for the lack of light and these are the high humidity and enhanced amounts of carbon dioxide. The plants down there have never had it so bad in terms of light, but could not have it better so far as humidity, shelter, more constant climate and the supply of carbon dioxide are concerned. Evolution

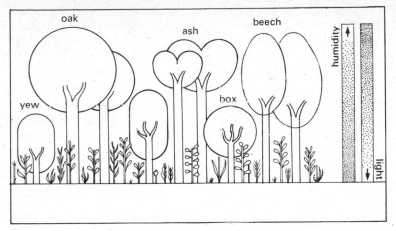

Fig. 36. Section of the mixed forest.

has fitted these shade plants to thrive in the conditions of the forest floor, and when one remembers that the seedlings of the canopy trees must be able to live and thrive under these conditions, the complexity of the process of evolution is obvious. A diversity of plants together can exploit the potential of the habitat with great efficiency.

Within the forest there is not only diversity in space, diversity in time is also built into the system. Early in the year the spring flowers reap the benefit of the sunlight filtering through the leafless branches – Windflower, Lesser Celandine among many others. Later the Bugle, Dogs Mercury, Bluebell and the forest grasses flower. Nothing is wasted and in the autumn, the fungi, toadstools and slime moulds (see chapter 6) live on the excess production of the year. The nutrients in the decaying leaves and eventually in the decaying fungi are remineralised and pass back to the ground helping to maintain the potential of the rich forest soil. In order to maintain a rich crop year after year, the farmer must till and fertilize his soil, rotating crops in order to keep the system productive. He will employ pesticides and fungicides to ward off epidemics which are a feature of crowded monocultures of any organism. In the complex of vegetation it is all built into the system. Darwin put it in a nut shell in his concept of simultaneous rotation, each member of the community doing its own job in harmony, each an integral part of a highly integrated system.

There are, however, at least two other types of forest on Box Hill, each less complex than the first and each more or less confined to its own special habitat. The mixed forest is found only on the soils developed over the layer of clay with flints which caps the hill. On the near vertical slopes, where no soil can collect, the woods are of Yew and Box which cling by the skin of their roots onto the bare chalk. Box, the tree which gives the hill its name, is not very much to look at but has a very tough, fine grained wood used for making rulers, and the Yew with its very strong, springy wood was the might of the English long bow. Below the cliffs, aptly named 'the Whites', still on almost bare chalk but with some soil enriched by erosion from above, there are the great Beech Hangers. Here are magnificent trees, soaring

Plate 27. Birds Nest Orchid.

skywards, producing a canopy of leaves high above the forest floor so dense that only thin pencils of light illuminate the gloom. The same thick tough leaves falling to the forest floor only decay very slowly and produce a raw acid humus. The forest floor is, therefore, a harsh place to live and few plants other than the toadstools manage to do so. Those that do include two strange members of the British flora, *Neottia nidus avis*, the Birds Nest Orchid (plate 27), and *Monotropa hypopitys*, the Yellow Birds Nest. Both have dropped the habit of photosynthesis and live like the toadstools on the products of decay of the forest floor. They are saprophytes and although they are certainly flowering plants, they do not contain the green pigment chlorophyll. At first sight both

Fig. 37. Butcher's Broom.

look very much alike, although a closer look soon shows that one is a member of the Orchid family, the ORCHIDACEAE, and the other is a member of the MONOTROPACEAE. This is a case of convergent evolution; two unrelated families of flowering plants each producing very similar products evolved to do the same type of job.

The best-known case of convergent evolution in the flowering plants are the two great families, the EUPHORBIACEAE and the CACTACEAE, both of which have produced cactus-like plants which exploit the harsh desert environment. The nearest thing to a cactus found in Britain is perhaps the Butchers Broom (fig 37) and it is found in the forest floor community. The plant consists of a mass of flattened branches which look and act like leaves each ending in a strong sharp spine. The flower and beautiful red berry is borne in the middle of the flattened stems in the axils of minute leaves. Butchers Broom, as its name suggests, was used for scrubbing down the butcher's block, as well as by masochistic sufferers from chilblains – believe it or not, to open the chilblains by beating their feet.

The forests of Box Hill hold some real surprises and none more so than the tangled masses of the stems of Old Man's Beard dangling down from the forest canopy like Lianas in a tropical forest. It is in good company, along with the Mistletoe and the Spurge Laurel, both of which are members of families which have their main homes down in the Tropics. Deep in the forest floor, conditions will be much more uniform throughout the year and if tropical-type plants are going to turn up in the temperate regions, they should occur here.

There is however, one thing which does appear out of place up on the hill and that is the famous grasslands which have drawn botanists and picnickers alike for many years. If the vegetation of the hill is not managed, why isn't it all under forest? Who mows the lawn?

When man first came on to the Surrey scene, he cleared the forests. The slopes of the hill were too steep for the plough and so they were used as pasture for the grazing of animals, especially sheep. The grasslands of the hill were first maintained by the nibbling of sheep. Later, as south-east man turned from the land to the commuter trains, the sheep walks fell into disuse. Perhaps it was the Victorian week-ender who did not like sitting on sheep dung, who knows, but the sheep went and if it had not been for the rabbits, the sheep walks would have been swallowed up by encroaching forest. Then myxomatosis swept the countryside and the lawn-mowing rabbits were gone; the fight was on as the process of succession to forest started with a vengeance. Now the only hope is a new brigade of 'rabbits', the Conservation Corps of the Council for Nature, whose trojan work is keeping the grasslands open.

One phenomenon still unexplained is the distribution of courting couples on the hill. The grasslands of the south-facing slope abound with such centres of biological activity, but not so the more extensive and more secluded north-facing slopes. The answer is in the microclimate. A unit of sunlight falling on the south-facing slope illuminates and warms only a small area of soil, whereas a similar unit shining just over the brow of the hill covers a much larger area with the same amount of heat energy. The south-facing slope is therefore hot and dry, the ideal place for human activities, whereas the north-facing slope is cool and damp, ideal for certain plants and supports a lush carpet of Mosses and even Liverworts.

This is vegetation: a pattern of ordered diversity, a product of the evolutionary process, a process which has covered the varied landscapes of the earth with self-maintaining productive communities of plants and animals, all working together and reaping the potential of the environment.

Do-It-Yourself Botany Kit

When I originally set out to make the programmes that go with this book, it was not my intention to produce a primer in botany. All I set out to do was to attempt to show the viewer/reader some of the things which excite me about the world of plants. The end result was rather a hotch potch of botanical fact which asked many more questions than it answered. A fact that was made abundantly clear by the enquiries which I received in their thousands.

The problem now is to take the whole thing back to square one and put some factual meat on the bones of interest. To this end I will try to highlight some of the important facts in each of the programmes by presenting more basic data. Each section of these notes will also list other books for you to peruse and/or things for you to do in your own backyard or on the kitchen table.

Chapter 1   Roots, Shoots and Flowers

Although grasses may be the commonest plants which we see in our gardens and much of our countryside, their leaves and flowers are, to say the least, somewhat specialised. So, in order to get down to basics how about taking a look at an ordinary plant growing in your garden. Go on, make the sacrifice, pull it up, a weed will do.

Here are a basic set of definitions of each part starting at the root of the matter.

ROOTS, are usually colourless (once you have washed the dirt off), they usually, but not always, grow underground. Their delicate growing tips are covered with a protective cap, like a little thimble, behind which you can often see a mass of delicate filaments, the root hairs which dry up very quickly on exposure to air. The main functions of roots are anchorage,   and   the   uptake   of   water   and   mineral   salts   from

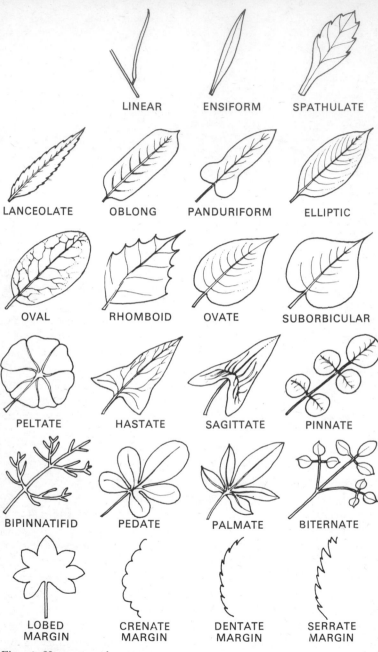

LINEAR ENSIFORM SPATHULATE

LANCEOLATE OBLONG PANDURIFORM ELLIPTIC

OVAL RHOMBOID OVATE SUBORBICULAR

PELTATE HASTATE SAGITTATE PINNATE

BIPINNATIFID PEDATE PALMATE BITERNATE

LOBED MARGIN CRENATE MARGIN DENTATE MARGIN SERRATE MARGIN

Fig. 38. Know your leaves.

solution in the soil. That's what all those hairs are for.

Now take a close look at a King Edward or come to that any less exalted common or garden spud. It is colourless and grows underground, but it does have buds and tiny scale leaves. If you don't believe your eyes then remember what happens when you plant a seed potato. A potato isn't a root, it is a specialised storage organ called a TUBER which is in fact a modified stem and the eyes of a potato are in fact buds.

STEMS, usually are green and grow above ground. Stems always bear buds in the axils of modified 'branches' called leaves, which are arranged at definite positions called nodes. The above ground part of most plants, that is the stems, leaves and flowers are referred to collectively as THE SHOOT.

BUDS are resting shoots that remain protected inside special scale leaves waiting for their turn to grow. Some buds are very small; take a look around your house plants. Some are very large; savour the fact next time you eat brussels sprouts. If your sprouts have lost their savour, dissect one carefully and see what its made of.

LEAVES, are usually flattened structures that are predominantly green in colour, the green of the chlorophyll may however be masked by other pigments, especially in autumn and, of course, in all those ornamental plants with flashy leaves. Leaves vary greatly in shape and size, but whether it is a minute scale, or a large compound thing like a horsechestnut, or a sharp spine as in some cacti, you will always find a bud in the axil of the leaf stalk or PETIOLE.

STIPULES. Apart from the bud in the axil, the base of the leaf stalk is often modified to form a stipule. Stipules range from broad leaf like expansions through to sharp stiff protective spines.

FLOWERS. At certain times of the year, and often for no apparent reason, shoots stop producing dull green leaves and switch over to producing brightly coloured flowers. This is not a haphazard process, but is carefully controlled by interaction with the genetic makeup of the plant and the environment.

BIENNIALS grow vegetatively for the first year, only flowering in the second year after they have been switched over by the cold of winter. Autumn flowering plants are switched on by the shorter days of late summer, while summer flowers require the long days of late spring. The changeover, though spectacular, is simply a modification of the normal growth pattern, because the SEPALS, PETALS and CARPELS which protect the embryo seeds within are all modified leaves.

Suggestions for Work

(1) Collect a leaf from each plant in your garden (or better still take your sketch pad into the garden), make a careful drawing of each one, labelling the parts. Where's the bud?

If you haven't got a garden, a rubbish dump, waste land or a quiet roadside verge will provide an abundance of material. Please whatever you do, don't go and pick flowers from our countryside, leave them there for everyone to enjoy and study. If you do have to stoop to study them in the wild, be very careful what you are stooping on.

(2) Keep a diary of what is in flower in your garden, weeds and all. Either get one of those plan your year diaries or rule up a piece of paper and stick it by the window that gives the best view of your garden, then every time you look out make a note.

(3) Using either a gardening catalogue or one of the books listed below, check up your garden plants and see to which family they belong. Take

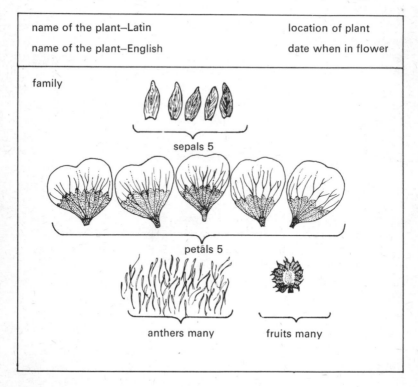

Fig. 39. File away your garden flowers.

one flower of each sort and carefully pull it to pieces, eyebrow tweezers will do. Arrange the pieces on a postcard and then stick them down with a piece of sellotape. They should keep very well. Fill in the details shown below, on the card. In a year you will be surprised what a fund of information you will have in your card index.

Further Reading

1. COLLINS POCKET GUIDE TO WILD FLOWERS. D. MCCLINTOCK, R. S. R. FITTER, Collins (1965).
This is a super book providing the beginner with an easy to understand guide to the British Flora.

2. PICTORIAL ENCYCLOPEDIA OF PLANTS AND FLOWERS. F. A. NOVAR. Paul Hamlyn (1965).
The book has got superb illustrations which will help you to find your way around the plant kingdom.

3. PENGUIN DICTIONARY OF BRITISH NATURAL HISTORY. R. S. R. FITTER, Penguin 1967.
A reference paperback covering most of the terms you will come across.

Chapter 2   Gathering Moss

I must confess that the two chapters on peat are somewhat specialised, they are in fact exactly what you might expect from a dedicated peat-nick like me. As you go about your country walks and especially while you are on holiday keep your eyes open for peatlands, they come in all shapes and degrees of wetness. The best way to get deeper acquaintance with these wet spots is to pay a visit to your local wetland nature reserve, where nature trails will guide you through without the problems of falling-in or inadvertently causing damage to the delicate wetland communities.

If you don't know where your local nature reserves are, then I suggest you contact and join your local Naturalist Trust, and if you don't know how to do that, then ask at your public library or citizens advice bureau. Same goes for your local natural history society. There are excellent societies in most parts of the country and if you really want to learn about your local plants and animals then join the local group and get into the field with the real experts.

Apart from the obscurities of peatland ecology, chapter 2 should have made you aware of the fact that one of the more important members of peatland flora are the mosses. This is also true of very many of the vegetation types which characterise these damp islands of ours. It therefore seems an appropriate place to introduce you to the mini world of this very successful group of plants.

Walls, rocks, tree trunks, waterfalls, grasslands, yes even the best kept lawn can abound with mosses, so long as it is damp.

Mosses are POIKILOHYDROUS, which means that they cannot grow up into dry air. This doesn't mean that they can't be found in dry places. Mosses are often abundant on very dry rocks, but in such habitats they spend most of their time 'all dried up' their life process just ticking over, waiting for the water supply to be restored (see fig. 10).

All the larger more complex plants, the ferns (PTERIDOPHYTES), conifers (GYMNOSPERMS) and the flowering plants (ANGIOSPERMS) can grow up into dry air, at least for as long as their roots are supplied with water. They are said to be HOMIOHYDROUS, because they can exercise a certain amount of control over loss of water. The reasons are complex, first their plant bodies are covered with a water-tight layer of fat called CUTIN, and secondly they all have a system of internal pipes (see chapter 5) which conducts water up from the roots keeping the shoot all alive oh! A collective name for all these HOMIOHYDROUS plants is VASCULAR, those with a built in plumbing system! Mosses do not have a system for internal water conduction and that is one reason why they wilt so quickly.

Go on, back into the garden and find a moss. Hold it up to the light and take a close look, a magnifying glass will help. There are the delicate leaves arranged in a spiral fashion around the stem. The fact that you can see the stem through the leaves emphasises their delicate nature. The leaves of all mosses are only one cell thick and although some are tougher than others, they all lose water very rapidly.

If you have got a good lens, exceptional eyesight or a large moss you may see that each leaf has a central vein or nerve. Don't be fooled, its main function is to give strength to the leaf, it cannot conduct water up from below because the nerve does not join up with any conducting tissue in the stem.

Put the mosses on your calendar diary. At certain times of the year you may find that your favourite moss plant has changed. The patch is covered with numerous tiny capsules each one situated on the top of a stiff-bristle like SETA. The capsules are usually green at first but as they ripen they turn brown, yellow or even red. These are organs of vegetative reproduction and when ripe they shed thousands of minute spores each of which is capable of growing into a new moss plant, just like the one which started the whole cycle off. (Fig. 40.)

The really interesting fact is that the moss plant itself bears tiny male and/or female sex organs. The latter contains an egg and the former shed many male cells which are motile and owing to their small size are able to actively swim through the surface film of water to the egg where they accomplish the process of fertilisation. It is impossible to see this unless you have a lot of patience and a good microscope. Nevertheless at certain times of the year it is happening under your very lawnmower, and to prove it the crop of capsules will soon lift their spores up into the dry air. The consummation of this mini sub-aqua marriage is a fertilised egg or ZYGOTE which grows to produce a capsule and so it goes on.

Mosses lead a double life, the reproductive phase of the life cycle is called the GAMETOPHYTE and that is usually with us throughout the year. The main multiplicatory phase is called the SPOROPHYTE and that is usually in evidence for a much shorter time.

Mosses are simple plants tied irrevocably to damp habitats because without a film of free water they cannot complete their life cycle.

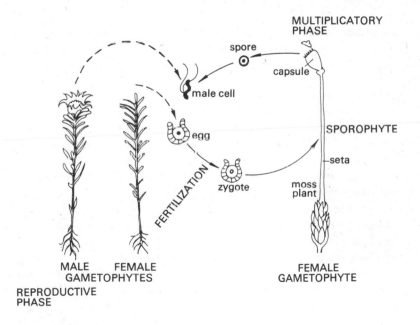

Fig. 40. Simplified life cycle of a moss.

Things to do

Keep your eyes open in your garden, and on your travels, yes even in the streets and note where the damper spots are, they will all be picked out by the presence of mosses.

Books to further your interest

OXFORD BOOK OF FLOWERLESS PLANTS. R. BRIGHTMAN, Oxford University Press.

This covers the mosses and a number of other plants you will also find growing in damp places.

BRITISH MOSSES AND LIVERWORTS. E. V. WATSON, Cambridge University Press (1968).

A very good guide to many of the mosses to be found in Britain, this is a book only for those who really get hooked on the group. This is not a book for rolling stones.

OBSERVERS BOOK OF MOSSES & LIVERWORTS. A. J. JEWELL, Warner & Co. Ltd. (1964).

A handy pocket book for beginners, who don't want to get too involved.

Chapter 3   Pollen and Spores

Take a close look at your card index of flowers. No, better still, go and get a fresh flower from the garden. Stick your nose well in and take a deep sniff, all those who suffer from hay fever do so at their own risk.

The stuff that sets you sneezing and/or the yellow stain on the end of your nose is pollen. Pollen is produced in the anthers, that is the male part of the flowers. Once released it is carried by insects, wind, or in very special cases by water, bats, birds or even human noses to another flower. Those pollen grains that make the journey and arrive on the female part of the flower (and it must be a flower of the right type and the stigma must be in a receptive state at the time), may bring about the process of fertilisation. The whole thing is of course a very chancy business and that is the main reason why most flowers produce an enormous amount of pollen 'bless you'.

Once the right pollen grain has got to the right stigma things can start to happen quite quickly. The tough coat of the pollen grain ruptures and a fine tube grows out into the tissue of the stigma, down towards the ovaries. On reaching an ovary the end of the pollen tube ruptures releasing two very reduced male cells which fertilise the egg. They cannot, and do not, swim to the egg and this is the secret of the flowering plants. Their up-tight botanical name of ANGIOSPERMS is derived from two Greek words meaning hidden or enclosed seed.

It is the protection afforded by the wall of the ovary that has allowed the flowering plants to kick the habit of a wet habitat. The angiosperms are without doubt the most successful of the land plants, even penetrating into the more desert regions of the world. The majority of the mosses and, to a lesser extent, the ferns and conifers are restricted to the wetter parts and the wetter habitats of the globe.

The minute pollen grains are not the only examples of protection in the dry habitat racket. The fertilised egg develops safe inside the protection of the ovary into an embryo contained within a tough coat or testa. The package, which is double wrapped to prevent water loss is a SEED, and another name for the Angiosperms are the seed plants. The outer wrapper, which may take on vast proportions, is the fruit and it is made from the tissues of the ovary itself.

Many seeds can remain dormant for a long period of time so that only when conditions are just right will the seed coat split and the delicate embryo start to grow into a new plant.

Have a close look at all the seeds you come across in your everyday diet. Rice, corn off the cob, peas, beans, and a vast range of nuts, enough to tickle the most fastidious palate. The others, oranges, lemons, apples, grapes etc. are enough to give you the pip, and although they may get in our gastronomic way they are of vital importance to the plant.

Make a careful drawing of all the seeds you eat, or nearly eat in the course of a week. Same goes for all the succulent wrappings, the fruits. All fruits should contain seeds, some like the pomegranate contain many more than their fair share, others like bananas and seedless satsumas have been bred minus their vital parts, just to please us humans. How many sorts of fruits can you find?, check the common ones off against the table, and next time you eat a blackberry with your mouth full of those gritty little seeds you will be able to say with aplomb, what a succulent collection of DRUPULES.

Drupules means tiny single seeded fruits, and you can say it with aplomb because a plum is a drupe, a large fruit with a single seed that is encased in a hard woody case.

Fruits not only add colour to our autumn landscapes and our gourmet diets, but also to the language of botany. When you have had enough of drawing those fruits and seeds, try growing them, you may be surprised what a warm, draught free, corner and a little bit of loving care can do. My greatest successes are date stones and a coconut, yes coconuts are seeds modified for dispersal by water and given a warm

1. *Simple Fruits* are formed from a single ovary which may contain one or many ovules, each of which becomes a seed.

1a *Dry Fruits* Outer covering of the seed remains dry.

1a(1) *Indehiscent* Contains one seed which is not set free from the fruit
- (a) Achene, e.g. Dandelion
- (b) Nut, e.g. Walnut

1a(2) *Dehiscent* Contains many seeds which are liberated from the fruit
- (a) Legume e.g. Pea (leave it on the plant and see it dehisce)
- (b) Follicle e.g. Kingcup
- (c) Capsule e.g. Poppy

1a(3) *Shizocarpic* (splitting), e.g. Nasturtium (fruit splits to release the seed, each seed remains covered with part of the fruit)

1b *Succulent* The outer parts of the fruit become juicy.

1b(1) *Drupe*, e.g. Plum (the inner part of the fruit becomes hard forming a 'stone' protecting the seed)

1b(2) *Berry* (The whole of the fruit, except the outer skin is fleshy, e.g. Tomato is a many seeded berry, and a date is a single seeded berry)

2. *Compound fruits* are collections of simple fruits
Collection of Achenes, e.g. Buttercup
Collection of Follicles, e.g. Delphinium
Collection of Drupes, e.g. Raspberry

3. *False fruits* are formed when parts of the flower other than the ovary help to make up part of the fruit.

Apple – the core is the ovary, the flesh is a swollen receptacle which originally held the ovary.

Strawberry – the fruits are in fact tiny achenes (the hard bits on the outside) which are stuck onto a swollen receptacle.

Rose Hip – the fruits (which make good itching powder) are achenes which are borne inside the swollen receptacle.

. Please note that this list is not exhaustive, but should cover the majority of the types you will come across.

Fig. 41. The fruits of the field.

spot and a lot of water they do germinate.

If, like me you are one of those people who can't bear to wait for the PLUMULE (the young shoot) to appear above the ground, then germinate your seeds as shown in fig. 42, the only problem is that you need a very big jam jar for a coconut.

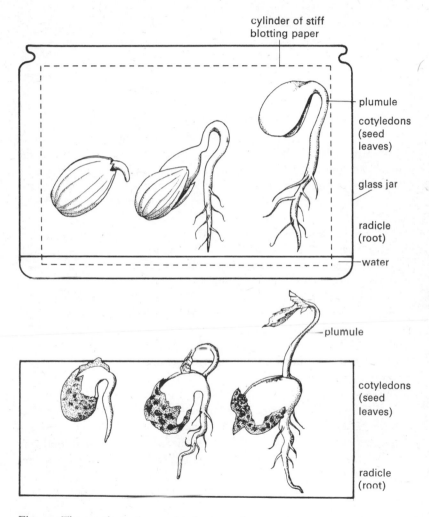

Fig. 42. The see through germination experiment.
Two types of germination, can you see the difference?

Chapter 4 Leaves Sold Only by Plants at all Branches

The umbrella like pattern which characterises the flowering heads (INFLORESCENCES) of the umbellifer family is just one of a whole range of branch patterns. Take a look around the garden or the local park and see how many you can find. At first these will appear to be a bewildering variety, stick with it because a closer look will reveal a set of basic patterns in what appears to be the disorder of branching.

The simplest and indeed most primitive form of branching is called DICHOTOMOUS. In this type, every branch point produces two equal branches. The best place to see this type of branching is down beside the seaside, in front of the Kingdom of Canute, where many of the seaweeds display this simple regularity.

From here on up the branch patterns get more and more complex. They can however be grouped into two basic forms, CYMOSE and RACEMOSE. Remember leaves are borne at regular positions on the stem, called nodes, and the buds which may eventually give rise to a new branch are borne in the axils of leaves.

Leaves may be arranged in pairs at each node, when they are termed OPPOSITE. Often each adjacent set is turned at right angles to the set immediately above and below when they are said to be OPPOSITE AND DECUSSATE. If more than two leaves are borne at each node then they are said to be WHORLED. In many cases you will see that only one leaf arises at each node and these are then termed ALTERNATE.

Find a plant with alternate leaves and tie a piece of thread onto one of the leaf stalks near its base. Wind the thread around the stem so that it touches the base of each leaf, carry on until you come to the next leaf that is immediately above the one you started from. You are now getting to grips with the very exacting science of leaf arrangement or PHYLLOTAXY.

If you invest in a lot of thread and carry on winding pieces around all the different plants you can lay your hands on, you will find that leaves can be 2, 3, 5, 8, 13, 21, 34 or 55 ranked. A 2 ranked leaf is easy, it means that every leaf on the spiral is directly above the second one below it. Double dutch? no check your thread. An 8 ranked arrangement means that any leaf will be directly above the eighth one below it, and a 55 ranked leaf, well you've got an awful lot of twiddling to do before you get to the next leaf in the vertical series.

The whole thing may be further complicated by the fact that the stem is itself twisted and then it takes a lot of sorting out. However with

sufficient thread and a lot of patience you should be able to unravel the
PHYLLOTAXIS of many of your garden plants.

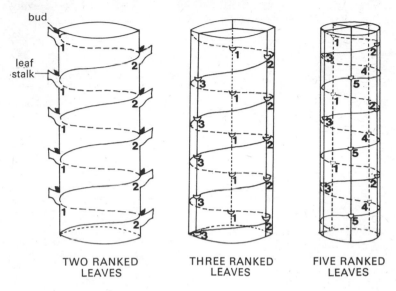

Fig. 43. The first three ranks of Fibbonachi.

The really fascinating thing is that the series of numbers 2, 3, 5, 8
and so on is not haphazard, it is in fact a strict mathematical progression
in which each figure is the sum of the preceding two, $2 + 3 = 5$;
$5 + 3 = 8$ and $21 + 34 = 55$. This series is called 'FIBBONACHI'
after the nickname of the mathematician who first discovered it.

Why all these leaf arrangements should obey the rules of Fibbonachi's
series we don't pretend to know. The fact is that they do, and in so
doing they are arranged so that each leaf gets its fair share of the sunlight incident on the plant. Not a bad thing when you remember that
the function of the leaves is to trap the energy of the sun so that it can
be stored as starch. The mathematics of plant growth may take an
awful lot of explaining but at least it makes functional sense.

If vegetative growth is that complex, you might expect that reproductive growth would be a nightmare. Remember that the main parts of the
flowers are modified leaves and that flowers are borne on branches
which develop from buds. Hold onto your Fibbonachi series and here
goes.

A solitary flower is easy, it just sits on top of a stem, terminating its growth. However in many cases the flowers are borne in complex heads called INFLORESCENCES.

When each flower in a head terminates each branch and further branching is from axillary buds, the flower head is said to be CYMOSE, see fig. 44.

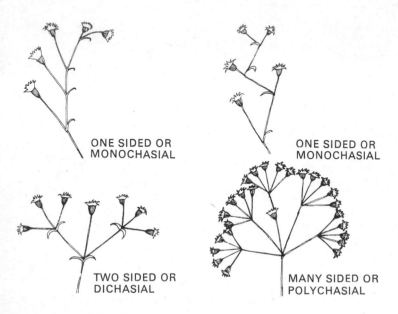

Fig. 44. The cymose family of inflorescences.

When the main stem goes on growing and the lateral branches bear the flowers, the branch pattern is said to be RACEMOSE, see fig. 45. The most complex are the open racemes or PANICLES of many grasses and the most difficult to study are the reduced racemes or SPIKES so typical of many other grasses.

A CORYMB is a special inflorescence which at first sight looks like the UMBEL of the Umbelliferae. Close inspection will, however, show you that the spokes in the cartwheel arise at different levels.

As each branch starts its life as a bud, each branch should be in the axil of a leaf. In many inflorescences these leaves are very much reduced, but however large or small they are called BRACTS. In some cases the bracts themselves are brightly coloured making the whole flower head more attractive to insects.

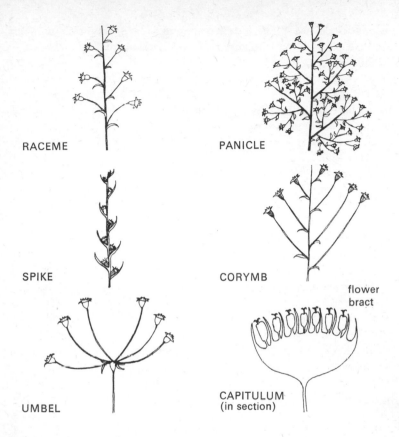

RACEME

PANICLE

SPIKE

CORYMB

UMBEL

flower
bract

CAPITULUM
(in section)

Fig. 45. The racemose family of inflorescences.

There are a number of flowers in which the branches of the flowering heads never elongate and a compact flowering head or CAPITULUM is produced. Such flower heads are features of the family the COMPOSITAE which includes such common garden weeds as the daisies and dandelions.

Take a daisy and see whether you can identify all the parts of the flower head, bracts, florets, anthers, stigmas and all. Now look at the arrangement of the tiny flowers on the head. They are arranged in two sets of interlocking spirals, one set running clockwise, the other anti-clockwise. If you count the flowers in each series you will find that there are 21 flowers in each of the clockwise spirals and 34 in each of the anticlockwise spirals. Twenty-one and 34, are Fibbonachi numbers; and I took up botany because I was no good at maths!

Books

THE SCIENCE OF BOTANY. PAUL WEISS, McGraw Hill (1962).
There are plenty of good modern texts of botany for you to choose from, far too many to list. I have chosen this one, simply because I like it and its a very nice book to own.

THE LANGUAGE OF MATHEMATICS. FRANK LAND, John Murray (1960).
If you weren't too good at maths at school, then get this book, it is compulsive reading and once digested it will allow you to look at many things, including plants in a new and fascinating way.

Chapter 5 Waterlogged

Don't just sit there and believe everything I have written, try it out for yourself. All you Constance Spry addicts, why not try a flower arrangement with a difference. Take a pint milk bottle, the ones with elegant tapering necks are ideal. Take the bottle plus a bucket of water out into the garden and find a convenient bush that you don't mind pruning. Quickly remove a leafy branch, and plunge the cut end into the bucket of water. Hold the bottle underwater and when full, thread the cut stem well down into the bottle leaving all the leaves sticking out of the top. Remove the bottle from the bucket and plug the neck with plasticine, making a couple of pin pricks through the plug to let the air in, mark the level of the water in the neck of the bottle, and there you have got it. A unique flower arrangement, guaranteed if not to grace your table at least to get the dinner guests asking questions.

Q. 'My dear . . . what is it?'

A. 'A transpirometer!'

and then you've got 'em, and probably yourself, because you will find that it can be great fun measuring the rate of uptake of the water. Stand your bottled plant in front of an electric fan, or better still one of those blow heaters and you can almost watch the water go. When you are fed up with your experiment, remove all the leaves and measure their area. This can be done by tracing their outline onto a piece of graph paper and counting up the squares. A little bit of leaf area can shift an awful lot of water.

If you want to enter the prize winning class of transpirometers then set up two, one with a leafy twig of a flowering plant, the other with a similar twig of a conifer. Now invite your local forestry commission man to dinner and advise him on the merits of planting the countryside with soft wood.

Hard and soft wood, now there's a problem, because some hard woods like balsa are very soft, and some soft woods like yew are very hard. These two terms are really terms of convenience. The former refers to the wood of the Angisperms which have both vessels and tracheides (see page 33) and the latter to the wood of the conifers or GYMNOSPERMS which only has tracheides.

GYMNOSPERM, means naked seed. Like the flowering plants the conifers produce pollen grains which are carried by the wind to the ovules borne in the female cones. Next time you are out in the country-side, collect a pine cone (off the ground, of course) and let it dry out so that it opens up. There lying on the cone scales are beautiful winged seeds, but note that they are not enclosed in tissue but lie naked on the surface of the scale. This was also true of the ovule prior to fertilisa-tion and hence the name naked seeds. Gymnosperms are seed plants but they are not quite as advanced in the protection racket as their more showy relatives, the plants with flowers.

Books

TRANSLOCATION. MICHAEL RICHARDSON, Edward Arnold (1975). This book will tell you all you need to know about how and what gets around in plants.

Chapter 6 Renaissance

I apologise for doing this, but here is another reminder about conservation. The more people who, like me, write books about plants, the more people will go out looking for them. The simple fact is that in the days of few people and lots of countryside a bunch of wild flowers didn't harm anyone. Today with lots of budding botanists and less countryside we must all refrain from picking plants. There are now laws about it and laws are made to be kept, not broken. So as you go about your business, botanical or otherwise, leave them growing where they are, and that goes for our fungi too.

Oh yes, I know the argument, most of a toadstool is under the ground, so we don't kill it if we pinch the fruiting body. True, but a photograph and/or a quick sketch is just as good and there may be another mycologist following you along and he doesn't want to go home disappointed.

There are however two novel ways of getting to close grips with the fungi. The first is to grow your own mushrooms, and you will find plenty of addresses from which to obtain information and material, in your gardening magazines. This method has the added wonder of succulent breakfasts.

The second is more bizarre and although it takes a lot of preparation, and clearing up afterwards, it's well worth the effort. Make yourself an old master, so old in fact that it is decaying.

First of all you need a corner of a room where your new art form can develop, undisturbed, some wooden battens, a packet of tacks and a large sheet of polythene. Construct the picture frame as shown in fig. 46,

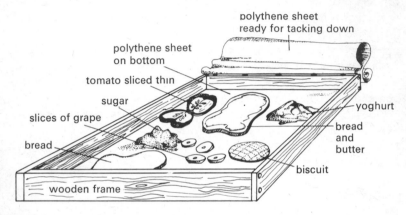

Fig. 46. The art of decay.

leaving the top sheet of polythene off. Cover the table with a sheet of polythene and then with a good layer of newspaper. Lay the picture frame closed side down on the paper. Now comes the creative bit. Take some slices of apples, pears, a grape or two, slices of bread (home made is the best), a dribble of yoghurt, in fact any organic material you can find and artistically arrange them inside the frame. Tack the polythene carefully on the top and then wait for it. Caution, the seal around the frame should be as air tight as possible, in order to keep both the moisture and the smell of decay in with the art nouveau. Sounds horrible doesn't it?, but it can produce some very beautiful effects.

As the organic matter begins to decay, bacteria and fungi grow in profusion. Only the expert could name all the microscopic plants which take up the palette of life, nevertheless you can still learn a lot from them. White, black, green, yellow and red colonies may expand in all directions. Some will overgrow others, while some will reign supreme pouring out antibiotics which inhibit all other growth in their vicinity.

Keep a diary of the events, better still make a photographic record of the changing pattern of life. Note how patterns of circulation of water build up, arising from one particularly lush bit of growth to trickle across the inside of the plastic sheet and fall into a pool of disintegrating yoghurt, entraining masses of dark green spores as it goes.

A word of warning, don't ever take the top off, the smell can be appalling. When you are fed up with your living picture, carry it carefully out into the garden and commend it, frame, polythene and all, to the bonfire.

I have a friend who got so attached to his that he kept it going for a year by which time it had dried out completely. A quick squirt of clear varnish and he had trapped it for ever, a collage of ever so still life.

Books

BIOLOGY OF THE FUNGI. C. T. INGOLD, Biological Monographs, Hutchinson (1961).

A book that has been around for a long time, but is still, for me, the best in the fungal field.

Chapter 7   Hearts of Oak

The might of the oak can be seen in two ways. First the almost proverbial seedling pushing up and appearing to lift any gravestones, which may bar its way to the light. Second, at the other end of its life span, a mighty tree with a symmetry of spreading branches, ideal for the main timbers of a ship of the non shivering wooden line. Between these two extremes is the whole wonder of growth. How does a soft green seedling ever grow up into a tapering woody giant?

We saw in chapter 1, that it is the apex which holds the key to the primary growth of most plants. The APICAL MERISTEM is an ordered mass of cells all of which have the ability of dividing and producing new cells. That is what the word meristem means – divisible. As the meristem grows forward it produces new tissue laying down initials which will in time produce all the cell types for life up in the dry air, or down in the damp soil, for both roots and shoots have apical meristems. Vessels and tracheides to carry the water, an outer skin or EPIDERMIS to protect the plant against water loss and attack by bacteria and fungi, the delicate packing tissue of the young stem and leaves, the cells of the stomata that allow ingress and egress of the gases so vital to the life of

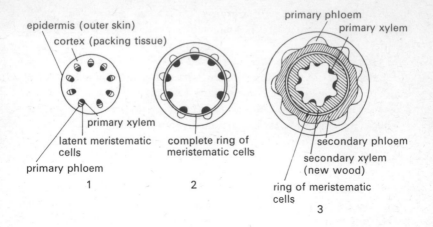

Fig. 47. The stages of secondary growth as seen under the microscope in a cross section of a stem.
1. Young stem with nine vascular bundles carrying water and sugar around the plant.
2. Latent meristematic cells have grown out to form a complete ring of meristematic cells.
3. Stem undergoing secondary thickening, the ring of meristematic cells is dividing to produce new xylem on the inside and new phloem toward the outside.

the plant. Many of the cells simply mature to their final form and maintain that form and function throughout their working life. Others situated at special positions within the packing tissue remain in a meristematic state, they change very little in form but their function of being able to divide and make new cells does not change.

As the young stem grows upwards it would soon begin to topple over (especially if the wind were blowing), unless a change in the pattern of growth took place. It does; those meristematic cells lying latent in the stem begin to divide producing new cells increasing the bulk of the stem providing the necessary support. New wood is produced towards the inside, an expanding plumbing system enough to keep the leaves well supplied with water. Around the new wood ordered masses of a soft living tissue the PHLOEM or BAST are produced in the same way. The main function of the phloem is to transport sugar produced in the leaves around the expanding plant body. The apical meristem which started the whole thing off is a PRIMARY MERISTEM which can only bring about growth in length. The SECONDARY MERISTEMS bring about growth in thickness, this secondary growth accounts for the transformation of a delicate green herb into a massive woody tree.

'Hold on, it can't be as simple as that, surely as the stem increases in girth the protective epidermis will split open.' It does, but already other secondary meristems closer to the surface have got cracking and dividing have produced a new protective layer, called CORK. As the tree grows in stature so more and more layers of cork are produced the whole forming the BARK, which is such a typical feature of most trees. It is indeed so typical that it can be used to help you identify the trees themselves.

Sally forth into the countryside and collect barks, but don't strip them off the trees, they really do need them. All you need is a piece of paper and a stick of artist's charcoal. Bark rubbing can be as much fun as brass rubbing and just as instructive.

Books

GUIDE TO BARK, ALFRED SCHWANKL, Open Air Guides. Thames and Hudson (1956).

Chapter 8   Denizens of the Shallows

Not long after Charles Darwin put forward his theory of evolution by natural selection it became dogmatised into what is no more than a catch phrase 'survival of the fittest'.

There is no getting away from the fact that the aquatic flowering plants have been fitted for life in the sweet waters of the world and do at least reign supreme in the shallows. This must be looked upon as a real story of success, especially when you take into consideration the fact that the angiosperms have only been around for a mere 150 million years. That may seem a long time to us mortals, it is however a very short period when viewed in relation to the 2000 plus million years during which evolution has been taking place.

If success is to be measured in terms of survival alone then the ALGAE which have graced the watery world for at least 1000 million years must be labelled as the true successes of plant evolution.

Algae, is a term of convenience which encompasses a whole range of plants, from minute unicells that are so small that only the electron microscope can reveal their patterns of life, through to the large seaweeds which help to make our sea shores places of great interest. Even the largest, and one of the kelps which grows on the coast of California can reach an incredible 200 ft in length, are relatively speaking simple plants. Their bodies are not divided up into stem roots and leaf, and are called THALLI, singular THALLUS. This is also true of the FUNGI, but there is one striking difference between the two groups. The majority of the algae contain chlorophyll and are photosynthetic, the fungi do not and are not.

Despite the overriding and all important presence of chlorophyll, the algae do come in a variety of colours. The other pigments which may or may not mask the all important green are used as diagnostic features of each algal group and so we have the following main divisions of these simple water plants. Blue green, green, golden, yellow-green, red and brown to name some of the most important groups. Don't run away with the idea that algae only grow in the sea; there are many fresh water forms and even some hardy thalli, which eke out an existence on land.

The vast bulk of the algae are too small to be studied without the aid of a good microscope. The majority of the large conspicuous ones which we see in the tide pools and washed up on the beach belong to the green, red and brown groups. It is these three groups that include all the plants which the layman would recognise as seaweeds.

The best way to study seaweeds is to invest in a copy of the local tide tables and mark in your diary all the best low tides. Then you can really take your interest down between the tides, which is the harsh home of many of our seaweeds.

Another way which doesn't necessitate getting up in the early hours to catch the tide is to go beach combing after stormy weather when with a bit of luck you will be able to find many different sorts of algae stranded on the beach.

Always bearing conservation in mind this is the time to collect them because once detached from their rocky perches they will die and rapidly decay away. Now comes the problem! Seaweeds are much harder to deal with than ordinary plants, the reason being that all the slime, which they so readily secrete, tends to dry and gets stuck to everything.

Have ready some sheets of good bond typing paper and some cheap butter muslin. Get your collection of flotsam ready beside a large bowl of water. Float each seaweed in turn in the bowl and slip a piece of the paper underneath the floating weed. Lift the paper carefully so that the weed floats out and arranges itself on the paper. A bit of gentle prodding may help the process. Now lay the paper together with its weed on a couple of sheets of newspaper and cover the specimen with a piece of the muslin, finally cover the whole with more sheets of newspaper. Float out another seaweed and repeat the whole process, piling one seaweed sandwich on top of the other. When all are complete, place some large books on top of the pile in order to press them. The important thing to remember is to change the newspaper, daily or even twice a day at first and then at longer intervals until the whole pile is bone dry. The pieces of muslin may now be carefully peeled off and with luck the seaweed will remain stuck to the paper, mounted for ever. Now add the name of the seaweed and the date and place you collected it and you have started an algal herbarium.

Failure to dry the seaweeds rapidly enough will result in another decaying masterpiece, success can give you the most beautiful and instructive record of your summer holiday. Please remember the date and location, your specimen could be of great importance as a record for the future, and please remember do not pull up attached seaweeds. Our coastlines are already being changed by promenades, pollution, oil slicks and all manner of man-made artefacts, do not hasten their despoliation.

Add to this the fact that in 1972 Japweed (*Sargassum muticum*) appeared for the first time in British waters. As its common name suggests, it is a seaweed of Japanese origin and was probably introduced to England via British Columbia. In a matter of only 30 years this particular seaweed spread along 2,500 km of the west coast of North America causing problems to boat owners and fishermen and undoubtedly changing the ecology of the area.

If Japweed invades the whole of the British coast, and despite vigorous attempts at eradication it is still on the rampage, your herbarium of seaweeds may well include this new invader in the not too far distant future.

We may well have another unwanted algal problem on our hands in the next few years. Some French industrialists want to experiment with the cultivation of the largest seaweed of all *Macrocystis pyrifera* along the French coast. The problem is that the inshore marine environment from Portugal to almost half way up Norway appears to be ideal for the growth of this plant. If the French do plant it, and it gets away we will probably be able to walk across the channel into Europe. No need for a tunnel or a referendum.

Books

THE BIOLOGY OF THE MARINE ALGAE. A. D. BONEY, Hutchinsons Biological Monographs (1969).
A super book giving you all the basic lowdown on our seaweeds.

Chapter 9   Buttercups and Floras

Go back through the book and underline all the terms we have used. Make sure that you know the meaning of each one. Now you are ready to have a bash at identifying an unknown plant.

The current, up to date, Flora of the British Isles is by three deservedly famous botanists, Sir Roy Clapham, Tom Tutin and the late Heff

Key to the British Buttercups with yellow flowers, that is to the Yellow Flowered Members of the Genus Ranunculus

	Sepals 3, Petals 7–12, Leaves simple	LESSER CELANDINE (*Ranunculus ficaria*)
	Sepals 5, Petals usually 5 leaves various	2
2	Leaves compound often palmate Leaves simple, Margins entire or toothed	3 11
3	Fruits covered with small lumps Fruits with spines or hooked hairs	4 10
4	Sepals strongly reflexed when flowering Sepals not strong reflexed	5 6
5	Stem Tuberous (Swollen) at the base Stem not Tuberous	BULBOUS BUTTERCUP (*R. bulbosus*) HAIRY BUTTERCUP (*R. sardous*)
6	Leaves hairless, or with few short hairs Leaves Hairy	7 8
7	Flowers 0.5–1 cm in diameter Flowers 1.5–2.5 cm in diameter	CELERY LEAVED BUTTERCUP (*R. sceleratus*) GOLDILOCKS (*R. auricomus*)
8	Plant with fleshy root tubers Plant without root tubers	FAN LEAVED BUTTERCUP (*R. flabellatus*) 9
9	Plant with long runners, peduncle furrowed Plant without long runners, peduncle not furrowed	CREEPING BUTTERCUP (*R. repens*) MEADOW BUTTERCUP (*R. acris*)
10	Flowers 3–6 mm diameter fruits covered with short hooks Flowers 4–12 mm diameter fruits spiny with the largest spines in the margin	SMALL FLOWERED BUTTERCUP (*R. parviflorus*) CORN CROWFOOT (*R. arvensis*)
11	Plant tall 60–90 cm flowers 2–3 cm diameter Plant less than 60 cm tall flowers less than 2 cm diameter	GREAT SPEARWORT (*R. lingua*) 12
12	Basal leaves more or less cordate, fruits with lumps Basal leaves not cordate, fruits smooth	SNAKESTONGUE CROWFOOT (*B. ophioglossifolius*) 13

	Flowers solitary 5–10 mm diameter fruit	LAKE CROWFOOT
13	1 mm long	(*R. reptans*)
	Flowers 1 – several 7–18 mm diameter fruits	LESSER SPEARWORT
	1.5–2.00mm long	(*R. flammula*)

Warburgh. It is called C.T.W. for short and comes in two sizes, the super de-luxe model containing exhaustive descriptions of all our vascular plants and the smaller put it in your pocket, eminently portable Excursion Flora. If you are really going to become an ever so British Botanist, then you must own a C.T.W.

However, before you rush out and invest in one, here is a chance to put yourself to the test, prove that you can identify a plant for yourself from scratch. Here is a simplified key to all the Buttercups with yellow flowers which you could find in your journeys around Britain. Go out and find a buttercup, you only need one plant, now follow the key and find out exactly which one you have got.

When you think that you have got to the right one, then check the shape of the fruits against those in the pictures. If the fruits do not fit then go back to the start of the key, do not pass go and do not collect a new buttercup, but try, try again.

The most important characters used in identifying plants are those which remain constant. As the reproductive parts are exposed to the environment for only a short time compared with the vegetative organs, they are usually much less variable. This is one reason why the fruits of the buttercups may be used in the final diagnosis.

Take a look around your garden buttercup patch and see how much the size of the plant and the size and shape of the leaves vary. Then measure up the flowers and see just how uniform they are.

The final check for any identification is to go to a full description of the plant and check your specimen against it in every detail. Descriptions of all our plants will be found in C.T.W., it's well worth the investment. The book will open up the world of the British Flora to you. I offer only two words of warning: 1) Don't try it out on the ornamental flowers in your garden, more often than not it just won't work, the reason is that most of them just aren't British. 2) If you enjoy it and get hooked, you're hooked for life.

Floras

EXCURSION FLORA OF THE BRITISH ISLES. CLAPHAM R., TUTIN, T. AND WARBURG, H., Cambridge (1959).
FLORA OF THE BRITISH ISLES. C.T.W., Cambridge (1957).

BRITISH FLORA. BENTHAM AND HOOKER, L. Reeve & Co.
(1924).
An older flora, still very good and can be picked up on the second-hand market.

Chapter 10   Form and Function

If you found the identification of the buttercup a bit of a drag, do not despair, you can still get a lot out of Botany without putting the exact names to plants.

For example, the GROWTH FORM of a plant can tell you a lot about its function within and relationship with the other members of the plant communities in which it grows.

Growth form, or LIFE FORM is another method of classifying plants, but it is a classification based on gross form and function.

(1) PHANAEROPHYTES – are plants which bear their buds more than 25 cms above soil level.

(2) CHAMAEPHYTES – are plants with their buds above the level of the soil but below 25 cms.

(3) HEMICRYPTOPHYTES – are plants which have their buds at soil level.

(4) GEOPHYTES – are plants which have their buds below soil level.

(5) THEROPHYTES – are plants which pass unfavourable conditions as seeds.

The trees and tall shrubs all of which would come under the term PHANAEROPHYTES, must have their buds well covered with tough scale leaves so that they can ride out the rigours of our winter up in the cold, dry wind. The buds of the CHAMAEPHYTES being much closer to the ground may be protected by a good fall of snow for the worst part of the winter. It is these dwarf shrubs which dominate our uplands where the winter is that much colder and the likelihood of snow that much greater.

The buds of the hemicryptophytes may remain dormant, snug beneath the snow and the litter of the previous year's fallen leaves. They, however, have another distinct advantage in that their buds are much more difficult to eat, the main problem being a mouthful of soil at each bite. The buds of the geophytes (those plants with underground storage organs) are even further protected from the vagaries of herbivores and lawn mowers. Down in the damp soil they are also protected from drying out during periods of drought and from the devastations of fires, be they natural or man made.

LESSER CELANDINE

BULBOUS BUTTERCUP

HAIRY BUTTERCUP

CELERY LEAVED BUTTERCUP

GOLDILOCKS

FAN LEAVED BUTTERCUP

CREEPING BUTTERCUP

MEADOW BUTTERCUP

SMALL FLOWERED BUTTERCUP

CORN CROWFOOT

GREAT SPEARWORT

SNAKESTONGUE CROWFOOT

LAKE CROWFOOT

LESSER SPEARWORT

Fig. 48. Do you like buttercup fruits?

Therophytes are the plants that we usually call annuals; for the simple reason that they grow only for one season in which they flower, fruit, seed and then die leaving only the seeds to carry on another generation. The term annual is misleading because many of them are able to produce more than one generation in a calendar year, responding to the shorter fluctuations of environmental change.

This simple life form classification was first used by a Danish botanist called Raunkier, as he tried to understand vegetation. As you go about Britain and even more so when in the strange alien climates of your European holidays, analyse the vegetation you see in terms of life form and record them as quick life form sketches.

If, like me, you are hopeless at drawing, then formalise each life form into specific shapes and draw sections of the vegetation like the one in fig. 36. These will form a unique record of the environments of your holiday. Here are some examples.

High up in the alps the vegetation will be dominated by CHAMAE-PHYTES whenever snow affords the right amount of winter protection. Wherever the snow cannot lie the communities will be dominated by therophytes. Similarly in the dry heat of the Mediterranean summer many plants will AESTIVATE (survive the drought) in the form of seeds or with underground food stores, germinating and bursting forth to flower in the rains of the following spring. It's all a matter of life form, and the motto of all plants is 'have the right life form, will function in the right environment'.

Add the new dimension of life form to your garden calendar and card index. In what sort of climate and type of vegetation did your choice blooms originate?

Books

FLOWERS OF EUROPE. POLUNIN O., Oxford Univ. Press. (1969).

Things You Can Do

The best way to get to grips with the subject of Botany is to get out into the field with other botanists and learn by contact. You can do this in one, two or three ways.

(1) Most areas of our countryside are now covered by local Naturalist Trusts. Although these are in the main bodies which deal with conservation, many of their activities can provide you with the all important contact, often in a very practical way. Also, as I have stressed throughout the book, conservation is the key concept in today's country-

side, so it is best to be led onto the straight and narrow, right from the start.

The same goes for your local natural history society. There are some very good ones with long and distinguished records of attracting and training our up and coming botanists.

To make contact with these local bodies ask for their addresses at your local museum, library, or citizens advice bureau.

(2) On a national scale there are two societies which specifically look after the plant side of things.

(a) The B.S.B.I. Botanical Society of the British Isles. If you are going to be a real British Botanist then you must join, write to:

Honorary General Secretary B.S.B.I.,
c/o Department of Botany,
British Museum (Natural History),
London SW7 5BD.

(b) The Wild Flower Society. This is a nationwide society whose members make it their job each year to check on exactly what is flowering in our countryside. Their regional and national activities have provided the first stimulus for a number of people who now rank amongst our best field botanists.

Wild Flower Society,
Rams Hill House, Tonbridge,
Kent.

(3) Sign on for a course.

(a) We are very fortunate in having the Field Studies Council with centres dotted in strategic points around Britain. Each year they run first rate courses both at the beginners and more advanced levels. Most of them last a week and are fully residential. They provide field training at its very best, write for details to:

The Field Studies Council, Preston Mountford Field Centre,
Preston Montford, Near Shrewsbury, Salop.

Field Study Centres

The Leonard Wills Field Centre,
Williton, Taunton, Somerset.
Flatford Mill Field Centre,
East Bergholt, Colchester.
Juniper Hall Field Centre,
Dorking, Surrey.
Malham Tarn Field Centre,
Nr. Settle, Yorkshire.

(b) Every University has its Extra Mural Department, and many of them run evening courses in the biological sciences. Don't be put off by the fact that they are labelled with the University ticket, many of them are designed for you, the beginner and they are taught by the top brass in the business. I suggest you take the plunge and enquire what's going on in your district. Write to:

The Director, Extra Mural Department, The University of So & So.

Finally, I still like scouring the local second-hand shops for botany books. There are some super ones and even if you don't get a bargain it can stimulate your interest.

Good botanising and please remember leave the wild plants exactly where they are for everyone to enjoy and learn about.